*f*P

Also by New Scientist

Stroke a Martian:
And 99 Other Things to Do Before You Die

100 Things to Do Before You Die
(Plus a Few to Do Afterwards)

Does Anything Eat Wasps?

And 101 Other Unsettling, Witty Answers to Questions You Never Thought You Wanted to Ask

NewScientist

FREE PRESS
New York London Toronto Sydney

*f*P
FREE PRESS
A Division of Simon & Schuster, Inc.
1230 Avenue of the Americas
New York, NY 10020

First Free Press trade paperback edition 2006

Originally published in Great Britian in 2005
by Profile Books Ltd.

Published by arrangement with Profile Books

FREE PRESS and colophon are trademarks
of Simon & Schuster, Inc.

For information about special discounts for bulk purchases,
please contact Simon & Schuster Special Sales:1-800-456-6798
or business@simonandschuster.com

DESIGNED BY ERICH HOBBING

Manufactured in the United States of America

1 3 5 7 9 10 8 6 4 2

Library of Congress Cataloging-in-Publication data is available.

ISBN-13: 978-0-7432-9726-4
ISBN-10: 0-7432-9726-1

Contents

Does Anything Eat Wasps?

Introduction

When, in 1994, *New Scientist* began publishing "The Last Word," the magazine's weekly column of everyday science questions and answers provided by readers, one of the editors asked how long we expected the column to run. Estimates ranged from 12 months to five years. "If," suggested one, "we reach 10 years, I'd be amazed. That would be more than 500 weeks of questions—there simply aren't that many out there."

Twelve years later you are reading *Does Anything Eat Wasps?,* a witty and sometimes esoteric compilation of questions and answers from a column that shows no sign of running out of steam. In the last five years alone, readers have told us how fat you have to be to become bulletproof, why dark alcoholic drinks cause heavier hangovers than light ones, how to judge the amount of rain in a cloud, and why eating green potatoes can be downright dangerous. And "The Last Word" has even spawned a research project and scientific paper in the journal *Physica A.* So taken were a group of researchers in Spain and the United States by our question asking why Tia Maria and cream interact so dramatically that they set about finding the answer. You can read what they discovered on p. 103.

So why has the column thrived against all expectations? Well, as colleagues often tell me, I have the easiest job on the magazine. "The Last Word" is driven entirely by the enthusiasm of *New Scientist*'s readers. Without their enduring input there would be no "Last Word," and you would not be reading this book. Every week our e-mail system is inundated with readers' new questions, and almost as rapidly, those

questions are answered by their peers. This book is a result of their efforts.

And, if you enjoy reading it, you can join them by buying the weekly magazine or logging on to http:/www. newscientist.com/lastword.ns, where you can pose your own question or answer another. But remember: "The Last Word" is devoted to the small questions in life. We can't solve the mystery behind the meaning of human existence, but we can tell you why your tea changes color when you add lemon juice. We don't know whether there is life in another galaxy, but we do know how to make bubbles in your chocolate bar. We are devoted to the trivial.

Back in 1994 that same skeptical editor promised to throw a party if "The Last Word" was still in existence in 2004. As well as waiting vainly for the column to show signs of flagging, I am still waiting for the party to which I should invite all those who have made "The Last Word" the success it is.

MICK O'HARE

The editor would like to extend special thanks to Jeremy Webb, Lucy Middleton, Alun Anderson, and the subediting and production teams of *New Scientist* magazine for helping to make this book so much better than it would have been otherwise.

Readers of the American edition of *New Scientist* will note that all measurements have been converted from the metric system, generally used in the magazine, to the measurements more widely used in the United States.

1 Our Bodies

Breaking Glass

My friend Paul and I can both hold a tune, but when he sings he sounds like Pavarotti, while I'm more like a wounded hippo. What are the anatomical prerequisites for a naturally pleasant singing voice?

CHRIS NEWTON

As with all musical instruments, the quality of the human voice is largely determined by resonance. A great Italian tenor gave permission for his larynx to be studied after his death. Air blown through his isolated vocal cords produced the same noise (somewhere between blowing a raspberry and the sound of a whoopee cushion) as in other, musically uncelebrated corpses.

The fundamental vibration of the vocal cord becomes operatic splendor by the addition of desirable harmonics. Some of these are imparted by immutable anatomy, such as the shape and density of the teeth, jaw, and sinuses, and others arise by manipulating the shape of the singer's oropharynx, the part of the throat at the back of the mouth. Fortunately the oropharynx can be trained and refined. Presumably those with a "naturally pleasant singing voice" have a lucky coincidence of skeleton and oropharyngeal shape.

DAVID WILLIAMS

Vocal sounds originate in the airstream that is forced through the larynx, the front of which is visible as the

Adam's apple, often prominent in males. Near the bottom of the larynx are the vocal folds, a pair of flaps that are essentially muscles whose thickness, area, shape, and tension can be controlled. The folds are open when you breathe, but when you produce a sound they come together and the air pressure builds up below them until they are forced apart. After reducing the pressure in a puff of air, the folds close again. Repeating the process creates a pulsating stream of air whose frequency determines the pitch of the sound.

Words, whether spoken or sung, are created by modifying this airstream. Most consonants chop it up into blocks whereas vowels are the sustained sounds in those blocks. The letters w, y, l, and r lie between these two extremes, constricting the airstream to modify its acoustic properties. After leaving the larynx, the airstream passes through the oropharynx—the upper part of the gullet—and into the mouth. Together, these structures can be likened to the tubing of a brass instrument between the mouthpiece (the vocal folds) and the bell (the lips). The air in it, like the air in any tube, has resonant frequencies, known as formants. In an adult man, the lowest formant is about 500 hertz. Changing the shape of the tube by arching the tongue, opening the jaws, modifying the shape of the lips, or altering the position of the larynx will either raise or lower the frequency of each formant. To a certain extent we do this unconsciously, but singers learn how to control these parts of their anatomy.

Singers also have another formant, which is thought to be caused by a standing wave set up in the short tube between the vocal folds and the point where the relatively narrow larynx joins the wider oropharynx. In acoustical terms there is an impedance mismatch at that junction, so part of the sound energy is reflected back toward the vocal folds. The effect is weak in normal speech because the tube is very short, and in untrained singers it can shorten still further because their larynx tends to rise as they try to sing louder and at a higher pitch. But in trained singers the larynx descends, lengthening the tube, and the effect becomes very

significant. It adds clarity and projection to a voice and is responsible for the "ring" or "bloom," which contributes to the voice's timbre. For these reasons it is called the singer's formant. In basses it is about 2,400 hertz, rising to 3,200 hertz or more in sopranos.

The pulsating airstream has a fundamental frequency and a whole series of harmonics. Their strengths are modified by the formants to create the vowels. The art of singing is to do everything any human does but while making a pleasing sound, and using it to create a melodic line with carefully controlled variations in volume to convey feeling and expression.

The technique required to become a fine singer is on a par with that required to become, say, a concert pianist. However, a voice is a relatively fragile instrument that sooner or later begins to deteriorate.

There are many reasons why people cannot sing well even if they can produce sounds that are in tune. The range of frequencies produced by their vocal folds may not match their formants—a bass voice may be allied to baritone formants, for example. Some people may not be able to achieve sufficient control over their larynx or tongue to manipulate their formants. In other cases the vibrations of the vocal folds might be irregular. Sometimes the vocal folds do not close correctly or they may be too dry to function properly: alcohol and cigarette smoke can dry the mucosa that normally keeps them well lubricated.

RICHARD HOLROYD

Contusion Confusion

Why do bruises go through a range of colors before they fade? I can see why they would be red or purple, but what accounts for the yellowish-green color? And why do they

often take a day or two to appear? Surely the damage and bleeding occur at the time of injury.

RICK ROSSI

A bruise occurs when small capillary blood vessels break under the skin. The hemoglobin in this leaked blood gives the bruise its classic red-purplish hue. The body then calls in white blood cells to repair the damage at the site of the injury, causing the red cells to break down. This causing produces the substances that are responsible for the color changes.

The breakdown products of hemoglobin are biliverdin, which is green; and then bilirubin, which is yellow. Later, the debris at the bruise site clears and the color fades.

It is the same process that disposes of red cells past their use-by date. White cells called macrophages break down defunct red cells in the spleen, liver, bone marrow, and other tissues. Bilirubin is taken up by the liver, where it is converted to bile and used in the digestion of food. It is bilirubin that helps to give feces their characteristic color.

CLAIRE ADAMS

The breakdown product of hemoglobin, bilirubin, is yellowish in color and is normally excreted from the body as a component of bile. Bile itself is secreted to help digest fat. This is very efficient recycling.

An accumulation of excess bilirubin in our body can occur in medical conditions such as hepatitis, giving the skin a yellow tinge also known as jaundice. One can sometimes observe this in some newborn babies.

Jaundiced skin will itch because bilirubin is an irritant, while bruises are tender to touch. Ultraviolet light helps in breaking down bilirubin and is also the treatment for jaundiced babies.

FRANKIE WONG

Bruises sometimes take a long time to appear because the damage can occur deep in the body tissues. The body under

the skin is not of course an amorphous mass—it has discrete muscles and organs, separated by planes of fibrous tissue (these can be seen clearly when we look at joints of meat from the butcher). When blood leaks from damaged vessels it is often prevented from reaching the skin's surface quickly by these planes of tissue, or it may simply take a while to diffuse through subcutaneous tissue.

The fibrous tissue sheaths also explain why a bruise occasionally appears some distance from the original impact—the leaking blood has tracked under the sheath and surfaces only where the fibrous tissue ends.

Stewart Lloyd
Consultant occupational physician

Congener Congeniality

I recently picked up a leaflet published by Health Scotland which said the darker the color of my alcoholic drink, the worse my hangover would be. Whisky, red wine, or brandy would lead to a worse morning after than would vodka or white wine, because the darker drinks contain something called congeners. After experimenting, I have to say this seems to be the case. But is it the congeners? If so, what are they and what do they do?

Richard King

Most people consume alcoholic drinks for their ethanol content. However, many such drinks also contain amounts of other biologically active compounds known as congeners. Congeners include complex organic molecules such as polyphenols, other alcohols such as methanol, and histamine. They are produced along with ethanol during fermentation or the drink's aging process.

Congeners are believed to contribute to the intoxicating effects of a drink, and the subsequent hangover. People who drink pure ethanol-based alcohols such as vodka have been shown to suffer fewer hangover symptoms than those who drink darker beverages such as whisky, brandy, and red wine, all of which have a much higher congener content.

The congener denounced as the main culprit in a hangover is methanol. Humans metabolize methanol in a similar way to ethanol, but the end product is different. Ethanol generates acetaldehyde, but when methanol is broken down, a major product is formaldehyde, which is more toxic than acetaldehyde and can cause blindness or death in high concentrations. Ethanol inhibits the metabolism of methanol; this may be why drinking "the hair of the dog" can alleviate hangover symptoms.

Studies have found that the severity of different drinks' hangover symptoms declines in this order: brandy, red wine, rum, whisky, white wine, gin, vodka, and pure ethanol.

ERIC ALBIE

Poison Pen

I've been reading a book called Deadly Doses: A Writer's Guide to Poisons *and it has raised two questions. When I was at school, my biology teacher told the class it was possible to drink snake venom and survive, because the venom is a protein and would be broken down by the digestive process. Yet the book says snake venom is deadly if taken orally. Who is right? And is it possible to build up a tolerance to arsenic, as in Dorothy L. Sayers's book* Strong Poison, *and so survive a lethal dose?*

DARREN FOWKES

Readers should be aware that there seems to be no consensus on what is safe. One thing is certain: these substances are dangerous and should be treated with extreme caution and not ingested in any form.—Ed.

I have watched a Zambian snake expert, Major Alick Chanda, milk the venom from a live puff adder (*Bitis arietans*) into a wineglass and drink it, with no ill effects.

Snake venom is a complex mixture of proteins, the composition varying from species to species. It can be classified into three major groups: cytotoxic, which attacks cells and tissues and is produced by vipers and adders; neurotoxic, which affects the nerves and is produced by cobras and mambas; and hemotoxic, which affects the blood-clotting process and is produced by back-fanged species such as the boomslang and bird snake. Whatever the type, the venom has to enter the bloodstream to have an effect. That is why snakes bite with hypodermic-like fangs. If you swallow venom, provided you have no lesions in your gastrointestinal tract, the proteins will be broken down into harmless amino acids and absorbed, like the products of all protein digestion.

Arsenic is an entirely different matter. Being an element, it is not affected by the digestive process. It is poisonous in doses of as little as 65 milligrams, and the poisoning can arise from a single large dose or from repeated small doses as, for example, by inhaling arsenical gases or dust, or drinking contaminated water.

There have been various accounts of people acquiring tolerance to poison by repeated small doses. In the second century B.C., King Mithridates IV was reputed to have used this method so successfully that when he tried to commit suicide by swallowing poison after defeat in the Battle of Pompey, he failed—and had to be stabbed to death. And Rasputin reportedly swallowed the stuff regularly to protect himself from being bumped off.

However, the physiological basis for such a tolerance has never been ascertained, and arsenic is known to be carcinogenic in small quantities. One has only to look at the effect of small repeated doses on the unfortunate inhabitants of Bangladesh, whose wells are contaminated with it, to see that this is harmful and ultimately fatal.

ALISTAIR SCOTT

Here in the United States, roadside and entertainment park exhibitions where snake handlers would handle venomous reptiles were once commonplace. In one of the most notorious, Ross Allen of Florida milked rattlesnakes into a glass and afterward drank the milk. He never exhibited any obvious ill effects, although he did say the venom made it "hard to whistle" for some time afterward.

Very high concentrations of venom given to animals orally, however, can have lethal consequences. The amount of venom administered for this experimental purpose is almost as much as the stomach can hold. If administered intravenously, these doses would be tens of thousands of times the LD50—that is, the amount of venom that kills 50 percent of a group of lab animals. When venom is given orally, death is the result of the venom itself and not a by-product of its digestion—administering antiserum saves the animal.

PROFESSOR JOSEPH F. GENNARO, JR. (RETIRED)
University of Florida College of Medicine, Gainesville

It is possible to build up a tolerance to arsenic and survive a dose that would normally be lethal.

Arsenic is toxic because it binds to, and thus inactivates, proteins that are essential to metabolism. However, it can be inactivated in the body by enzymes called metallothionines, and the presence of arsenic can induce liver cells to produce more of these. If small quantities of arsenic are consumed over a period of time, then enzyme production will be induced more often and background levels will increase.

This is the same sort of mechanism that allows alcoholics to consume quantities that would kill a teetotaller.

CRAIG FITZPATRICK

Only a fool would drink venom, because dangerous doses can get into the circulation through minor injuries to the mucous membrane of the digestive tract.

Many venoms contain enzymes that damage tissues and let poison in, and though toxic proteins certainly are deadlier when injected, some do pass through the gut wall. This factor is more important in infants, but even adults can die of swallowing small doses of very poisonous proteins, such as ricin. Gut enzymes break down some of the most dangerous venoms too slowly to give much protection, and many venoms resist or inhibit some of our digestive proteases.

As for inorganic arsenic, claims of tolerance are confused by travelers' tales, quackery, ill-defined preparations, and personal variation. Chemistry affects toxicity (arsenites are more poisonous than, say, arsenates or sulfur compounds) and the gut absorbs coarse-textured, poorly soluble material less easily. Soluble arsenic, far from inducing tolerance in small doses, is dangerously cumulative.

Dorothy L. Sayers's jam omelet story was cute, but before a would-be murderer had managed to kill anyone by this method, he would probably have expired.

JON RICHFIELD

High Brow

Why do people have eyebrows?

BEN HOLMES

My father has alopecia, so he doesn't have eyebrows. In warm weather, sweat runs into his eyes and makes them sore; in wet weather he has to keep wiping the rain out of his eyes. So your eyebrows divert sweat droplets and raindrops from running directly into your eyes. You would be very uncomfortable without them.

VALERIE HIGGINS

We use our exceptionally mobile eyebrows to communicate our emotions. The position of the eyebrows emphasizes expressions on the human face, thus giving others an accurate picture of the individual's mood. This gives a good indication of whether people are friendly or whether they might be dangerous to approach.

Smiles come in many forms, from expressions of merriment or contentment to leers, smirks, and even anger. The position of the brow, emphasized by the eyebrows, is what gives us a visual cue to what an individual is really feeling.

The importance of eyebrow position as a guide to mood was brought home to me when a friend had Botox injections in the lines on her forehead and couldn't raise or lower her eyebrows. Talking to her became a disconcerting experience—the bottom half of her face remained mobile but her eyebrows did not move. I couldn't deduce her mood accurately by looking at her expression, and needed to use other cues such as her actions and speech.

ALISON VENUGOBAN

Eyebrows are important in expressing emotions. Perhaps most important is the "eyebrow flash," a rapid up-and-down flick of the eyebrows that conveys recognition and approval. The ability to telegraph friendly intentions from a safe distance would have had obvious survival value for our ancestors.

Eyebrow signaling of various kinds is widespread among primates, although only in humans are the eyebrows highlighted by being set against bare skin.—Ed.

Life in a Glass

How long can human beings live if their sole source of food or drink is beer? And do different beers—ale, lager, stout, mild—confer a better chance of survival?

JOHN EDEN

Beer has had a reputation since antiquity as being a staple in the diet, often called "liquid bread." In ancient Egypt, workers received beer as part of their salary, as did the ladies-in-waiting of Queen Elizabeth I of England. In 1492, one gallon of beer per day was the standard allocation for sailors in the navy of Henry VII.

This high reputation for beer came about because it was made from malted barley, which is rich in vitamins. This is still true today. A quick check using nutritional tables shows that a pint can provide more than 5 percent of the daily recommended intake of several vitamins, such as B9, B6, and B2, although other vitamins such as A, C, and D are lacking.

It is of course unethical to conduct an experiment to see whether one can live on beer alone. However, during the Seven Years' War of 1756–63, John Clephane, physician to the English fleet, conducted a clinical trial. Three ships were sent from England to America. One—the *Grampus*—was supplied with plenty of beer, while the two control ships—the *Daedalus* and the *Tortoise*—had only the common allowance of spirits. After an unusually long voyage due to bad weather, Clephane reported that the *Daedalus* and *Tortoise* had 112 and 62 men respectively requiring hospitalization. The *Grampus*, on the other hand, had only 13, arguably a clear-cut result.

Needless to say, the sailors' allowance of 8 pints of beer per day is no longer within the accepted confines of current moderate alcohol consumption. One can only speculate on

the state of their livers. Living on beer alone may be a fantasy for some, but it is not a good health strategy.

C. WALKER
Brewing Research International, Nutfield, Surrey, UK

I do not know how long a human could survive on beer alone, although I suspect that the critical factors would be cirrhosis of the liver and vitamin deficiencies. Nonetheless, monks would be the ones to ask. Bavarian-style bock beers (or dark lagers) have for centuries been closely associated with monasteries, where they were brewed for times of fast and Lent. They are commonly known as liquid breads. Possibly the most famous beer of this style is Paulaner Salvator, and at 7.5 percent alcohol by volume it should keep your mind from food.

NEIL WATT

I offer the following answer: I'm 39 and still alive.

CHRIS JACK

I once put myself on a beer and cabbage diet. I can vouch that I lost weight, friends, and control of my lower bowel, often simultaneously.

BILL COPPOCK

Blubber Bullets

How fat would you have to be to be bulletproof, so that your fat layer would prevent a bullet fired from an ordinary handgun from reaching your vital organs? I recently read it was about 1,100 pounds, but find this hard to believe.

WARD VAN NOSTROM

The damage a bullet does to its target is measured in two ways: the depth of penetration and the amount of tissue damage per inch of penetration. These two figures are normally found by firing live rounds into blocks of a thick, viscous gel that is formulated to have the same physical properties, such as viscosity and density, as human flesh.

A 9-millimeter handgun round—the most common type—is quoted in the *Compendium of Modern Firearms* by K. Dockery and R. Talsorian (Games, 1991) as being able to penetrate approximately 24 inches of human flesh before it stops, doing an average of 0.06 cubic inch of damage per inch of penetration. In reality the distance penetrated is often much less, because rounds frequently hit bones or simply pass through the target. These data are also based on a general body tissue average. Because fat is approximately 10 percent softer and less dense than muscle, the figure of 24 inches may be too little.

Although being bulletproof may sound advantageous, carrying a 2-foot-thick layer of body fat obviously comes with its own health hazards.

THOMAS LAMBERT

A human body would never be entirely bulletproof when you take into consideration tissues and appendages such as the hands, feet, eyes, ears, and male genitals. Even if the skin was sufficiently thick to stop a bullet, the shock wave could seriously harm internal organs and the network of nerves below the skin, an effect which shot pellets exploit. Pellets from a shotgun can kill a human without penetrating the skin.

A bullet's depth of penetration in a body depends on a number of factors, such as the bullet's energy, diameter, mass, shape, and material. Bullets from rifles and handguns may range from approximately 5 to 15 millimeters in diameter and from 70 to 7,000 joules in energy. A typical police handgun bullet has a diameter of 9 millimeters and an initial energy of 500 joules. Penetration depth is measured in a

gelatin block, and the police handgun bullet typically penetrates about 12 inches of gelatin at a distance of 15 feet from the barrel.

To estimate how much such a fat layer would weigh, start with the surface area of the person "underneath." There are several formulas to calculate the body surface area; I will use the Mosteller formula, which gives an individual's body surface area. For a man 16 feet tall and weighing 165 pounds, this yields a body surface area of 21 square feet. So in order to cover this area with a 12-inch-thick layer of fat with a density of 0.6 ounce per cubic inch, we would need at least 1,260 pounds. When you add this to the weight of the body, you find that a typical bulletproof person would weigh about 1,425 pounds.

HANS-ULRICH MAST

Fossil Record

After my death I would like to become a fossil. Is there anything I could have done to my remains that would improve my chances, and where would be a good place to have them interred? How quickly could I turn into a fossil?

D. J. THOMPSON

So you want to become a fossil? This is admirable, but you have made a bad start. A hard, mineralized exoskeleton and a marine lifestyle would have given you a better chance. But let's start with what you have got: an internal skeleton and some soft outer bits.

You can usually forget the soft bits. If you take up mountaineering or skiing and end up in a glacier crevasse you could become a wizened mummy, but that's not real fossilization, just putting things on hold for a while. If you really want to survive the ravages of geological time, then you need to concentrate on teeth and bones. Fossilization of these involves additional mineralization, so you might want to get a head start and think about your diet: cheese and milk would build up your bone calcium. And look after your teeth, as these really are your best bet for a long-term future. So get a good dentist and keep those appointments.

After that it comes down to three things: location, location, location. You must find a place to die where you won't be disturbed for a long time. Caves have worked well for some people, so you might want to take up potholing to scout out locations close to home, but get the proper training.

Alternatively, you need a rapid burial. I don't mean a speedy funeral service by an undertaker picked out of the telephone directory, but something natural and dramatic—the sort of thing that is preceded by a distant volcanic rumble and an unfinished query along the lines of "What was . . . ?"

You might want to travel to find the right natural opportunity. Camping in a desert wadi in the flash-flood season would be good. And long walks across tropical river floodplains during heavy rain could get you where you want to be: buried in fine, anoxic mud. Or how about an imprudent picnic on the flanks of an active volcano? But take geological advice because you are looking for a nice ash-fall burial, not cremation by lava.

Talking of picnics, fossil stomach contents can provide useful paleo-diet information, so a solid final meal would be good. And I mean solid. Pizzas or hamburgers won't last, but shellfish or fruit with large seeds (you will need to swallow these) could intrigue future scientists.

Finally, trace fossils (marks in rocks that indicate animal behavior) are always welcome. So a neat set of footprints leading to your final location would be good. Use a nice even gait with no hopping or skipping to confuse analysis of how you really moved.

Of course, you have more chance of winning the lottery

than ending up as a fossil. But if you do go for a place in the fossil record please keep in touch. Geologists are always on the lookout for interesting new specimens, so let us know where you'll be. We can arrange to dig you up in, say, a million years.

TONY WEIGHELL

I learned the answer to this question 50 years ago, while studying geology at the University of St. Andrews in Scotland, but the name of the poet who explained it and the title escape me. I am quoting the text below from memory, and may have a few words wrong.

DAVE DUNCAN
Calgary, Alberta, Canada

Elegy Intended for Professor Buckland
Where shall we our great Professor inter,
That in peace may rest his bones?
If we hew him a rocky sepulchre,
He'll rise and break the stones,
And examine each stratum that lies around
For he is quite in his element underground.
If with mattock and spade his body we lay
In the common alluvial soil,
He'll start up and snatch those tools away,
Of his own geological toil;
In a stratum so young the Professor disdains
That embedded should lie his organic remains.
Then exposed to the drip of some case-hardening spring,
His carcase let stalactite cover,
And to Oxford the petrified sage let us bring,
When he is incrusted all over;
There, 'mid mammoths and crocodiles, high on a shelf,
Let him stand as a monument raised to himself.

We had to correct only a few of the words in the poem that Dave Duncan recalled from 50 years ago. This advice to let stalactites do the work of an embalmer is found in the second half of "Elegy Intended for Professor Buckland," written in 1820 by Richard Whately. William Buckland (1784–1856) held posts at the University of Oxford and was one of the most famous geologists of his day. He was a noted eccentric who claimed to have eaten his way through the entire animal kingdom. His contemporary Augustus Hare records how Buckland came across a casket in which the heart of a dead French king had been preserved: "He exclaimed, 'I have eaten many strange things, but have never eaten the heart of a king before,' and, before anyone could hinder him, he had gobbled it up, and the precious relic was lost for ever."—Ed.

Your chances of being fossilized are very slim, but you can significantly improve them by having your body buried at sea. However, you must make sure the water is very deep. This is because shallow marine conditions are turbulent and full of life, which will be only too glad to eat your remains. Terrestrial settings are liable to erosion, even if your body is buried, and therefore will significantly reduce your chance of fossilization. However, in the deep sea there are few creatures, and there are even fewer beneath the seabed, should you be able to have yourself interred there. Make sure that your chosen location is not near a tectonic subduction zone, where the Earth's crust is being consumed, because you will be quickly carried into the magma along with it.

The fine clay will help to preserve your body structure, and fossilization should proceed until you are nothing more than an outline of carbon and petrified body fluids, thanks mainly to compaction from the weight of clay that settles above you. You should allow around 200,000 years for this.

Of course, having yourself incarcerated in amber would

provide the best means of preservation, but you would then have to arrange to have the amber buried in a stable environment, which is not easy from beyond the grave.

Finally, wear something gold with your name on it, so you can be identified. It will probably survive far longer than your remains.

JON NOAD
Shell International Exploration Team, Rijswijk, Netherlands

Thanks to Oxford University Museum of Natural History, which provides a fuller story of Buckland's life at www.oum.ox.ac.uk/geocolls/buckland/bio1.htm.—Ed.

Delayed Reaction

Having just completed a half-marathon in the UK's Great North Run on Tyneside, I was surprised to find that my legs felt stiffer two days after the event than they were the following day. Why was this?

RUBY GOULD

Running is a form of eccentric exercise, meaning that the muscle is forced to lengthen while trying to contract. Prolonged or unaccustomed eccentric exercise often leads to pain, tenderness, and stiffness in the muscles hours or even days later. This is known as delayed onset muscle soreness and is very common.

The sensation of discomfort usually develops approximately 24 hours after exercise, peaks at about two days, and then gradually subsides. During the 24 to 48 hours post-exercise period, muscle swelling and stiffness usually result in a reduced range of motion and also muscle weakness.

Because the onset of muscle soreness is delayed, it cannot be attributed to the metabolic end products of exercise. In fact it is caused by localized damage such as microscopic tears to the membranes and protein filaments of muscle fibers. One hypothesis is that the damaged muscle cells die because they are subject to excessive calcium inflow. Another is that exercise-generated free radicals attack the cell membranes, and this attack leads to their death.

In addition, there is increased blood flow to the muscles, which causes the tissue to swell. Such swelling increases pressure on the neighboring structures. The nerves in the muscle sense this and send pain messages to your brain as you move the morning after you have exercised.

MELANIE TRICKETT

Delayed onset muscle soreness (DOMS) is a result of an excessive amount of tearing in the muscle. In order to improve performance when we exercise we need to progressively challenge our muscles with the amount of work we expect them to perform. This progressive overloading (usually achieved by increasing the resistance they experience such as by lifting heavier weights or by running extra distances on successive days) causes tears in the muscle's microfibers. And, in a gradual overload/repair cycle, we experience moderate soreness up to a day later.

DOMS is caused when the expected load dramatically increases, causing a greater number of tears (rather than an increase in the magnitude of each tear). In this situation it takes longer for scar tissue to form because it grows in perpendicular fashion across the repair sites. Once the new tissue is in place, we experience the soreness that comes with DOMS as we reactivate and stretch this new, less pliable muscle, until its strength and flexibility are restored.

PAUL CAREY
Personal trainer

The Sandman Cometh

My colleagues and I have been wondering, during those rare moments of reflection, what the scientific term is for the yellowish crystalline substance sometimes found encrusted on eyelids when you wake up. Some people call it "sand" or "sleep," but does it have a medical name? What is its composition and why does it form?

SIMON SMITH

The substance collects around the eyes because of irritation. During the day, the dried mucus consists of salts and proteins secreted by glands in response to dryness or exposure to pollution. The mucus continues to collect and dries out in the corners of your eyes while you're asleep even though tears keep the eyes moist.

The tears have three separate components. The innermost tear layer coats the surface of the cornea and is called the mucous layer or mucin. The middle tear layer is an aqueous layer produced by the lachrymal glands and supplies salt, proteins, and other compounds to the cornea. The outer tear layer is composed of oil from the meibomian sebaceous glands in the eyelids. This helps to prevent evaporation of the watery tears from the surface of the eyes.

Readers should be aware, however, that thick, frequent yellow or green mucus in the eyes is a sign of viral or bacterial conjunctivitis.

JOHAN UYS

There seems to be no widely used specific term, perhaps because the effect is seen as trivial and erratic. Nonetheless, it is important. During the day grit, dead cells and other debris accumulate in the tears which are more than mere salt water.

Mucoproteins cover the eyeball, curdling protectively around sharp grit to encase it in mucus; a middle salty layer is the main liquid part, and an outer, oily layer reduces evaporation. At night, movements of the eye and closed eyelids stir this orbital midden, massaging solids toward the inner corner of the eyelids. There the exposed liquid evaporates until the residual sludge forms pellets that you remove harmlessly by washing, or with your finger, the next morning.

Gritty environments such as deserts may damage eye tissues enough to convert your tears into dilute pus. This dries on the edges of your eyelids, gluing them shut in spite of the waxy coating that normally reduces spillage and keeps their epidermis water-repellent. It can be very disconcerting to awaken from an exhausted sleep to find your eyelids sealed shut so that you think it is still dark. If this ever happens to you, soak them open gently, or you may lose some eyelashes in the sand.

JON RICHFIELD

Neither of the above letters suggests a technical term for the substance that collects in the corner of your eye. Many terms were put forward but the most suitable we found was "mucopurulent discharge," offered by John Devers of Bronia, Victoria, Australia.—Ed.

Growth Areas

My doctor tells me that the fungus that causes athlete's foot tends to occur between the third and fourth toes. What is it about this area that the fungus prefers and how does this site differ from the space between all the other toes?

MARJORIE MCCLURE

Sadly, I am one of those unfortunate people who suffer from recurrent athlete's foot only between my third and fourth toes. All my other toes have a clear gap between them at their bases, but my third and fourth toes adjoin closely. This reduces the evaporation of moisture from this site, in turn creating a welcoming environment for fungus especially when I wear the same socks for 36 hours. Sorry, but this has been known when I have had to work all night on call.

In fact, I now prevent the return of athlete's foot by placing cotton balls or gauze between my third and fourth toes. This helps keep the area dry and is much cheaper than antifungal creams. Despite my having a medical doctorate and a PhD from Johns Hopkins University, my wife thinks that this treatment is half-baked.

John Criscione
Department of Biomedical Engineering
Texas A&M University, College Station

The organism responsible for tinea (better known as ringworm), or athlete's foot, does not have an intelligent territorial instinct that leads it to a predestined home. Infections by the fungus responsible, *Trichophyton mentagrophytes*, begin in the space between the third and fourth toes because this location offers the ideal environment: a warm, dark, and moist collection of dead skin cells.

The outer edges of the human foot are relatively flexible, having joints capable of motion in three planes. The spaces between all the other toes are therefore subject to a greater diversity of movement and forces, providing ventilation and the opportunity to slough off dead skin cells. Meanwhile, in the dark recesses between the less mobile third and fourth toes, a virtual agar plate awaits the arrival of *T. mentagrophytes*.

Felicity Prentice

Waxing Lyrical

What affects the different shadings of earwax? Sometimes mine is a light honey color; on other days it is very dark orange/brown. And why does its consistency change?

TONY COLUMBINE

About 2,000 sebum-producing sebaceous glands and specialized sweat-like apocrine glands in the outer third of the ear canal produce mildly acidic secretions, called cerumen. Earwax is a mixture of cerumen, skin cells, and hair fragments from the ear canal, plus bacteria and other substances caught in this waxy matrix.

Earwax normally moves out of the ear, and while there is some controversy over whether cerumen is bactericidal, the waxy substance is an effective material for catching dust, small particles, bacteria, and fungi that enter the ear. Earwax also lubricates the ear canal and has had a number of other uses because of its properties. It has even been used as lip balm.

There are two types of cerumen, dry and wet, the latter being the genetically dominant form. Both are controlled by a single autosomal gene. This has been used by anthropologists to trace migrations, because earwax in people of Mongolian Asian ancestry is usually the less common, recessive, dry type.

Cerumen is composed of glycerides, lipids—including squalene, cholesterol, and long-chain fatty acids—waxy esters, aromatic hydrocarbons, amino acids, and sugars such as galactose. Earwax also contains a complex of biochemicals from skin and hair, including significant amounts of collagen and keratin, and dead bacteria and fungi. Differences in the composition of these substances are one of the reasons that earwax builds up more in some people than in others.

Earwax color is a result of the light-absorbing properties of its chemical constituents. Wet and dry cerumen differ in lipid content: lipids account for about 50 percent in the former and 20 percent in the latter. So while dry earwax is somewhat crusty and typically grayish in color, relatively clean, fresh, wet earwax is typically a light-brownish honey color.

But the constituents also change with time. The color darkens because much of the long-chain fatty-acid content is unsaturated and slowly oxidizes on exposure to air. This yields a darker brown color. Eventual inclusion of dirt, dead cells, bacterial products, and hair fragments can turn earwax to an almost dark brown or black color in ears that are not cleaned frequently.

The initial color also varies due to idiosyncratic differences in the type and amount of glandular secretions that generate it. This balance can change in response to stress in the same way that the composition of sweat secretions does. Such changes reflect differences in the proportion of secretions from the sebaceous and apocrine glands, as well as variations in the concentration of the components.

Finally, as we age, even wet cerumen secretions become less liquid.

MARK DUBIN
University of Colorado, Boulder

Dead End

I've just had my appendix removed. My surgeon told me I wouldn't miss it because it no longer serves a purpose in humans. But does it serve a purpose in some animals? What exactly?

PAUL WHITTEN

The equivalent of the true appendix in most animals is known as the cecum and is at the junction of the small and large intestines. In general, carnivorous mammals have a small cecum that serves the same purpose as it does in humans. However, in many herbivorous mammals the cecum is greatly enlarged to create all sorts of wonderful anatomical arrangements. The function of the cecum in these mammals is to ferment the complex carbohydrates from the herbivorous diet into volatile fatty acids and then to absorb these as a source of energy. A functioning cecum is also vital for providing the energy needs of hindgut fermenters like horses, rabbits, rats, guinea pigs, and swine. The ruminant stomach of cattle and sheep performs a similar function in these animals and so they are less dependent on their ceca.

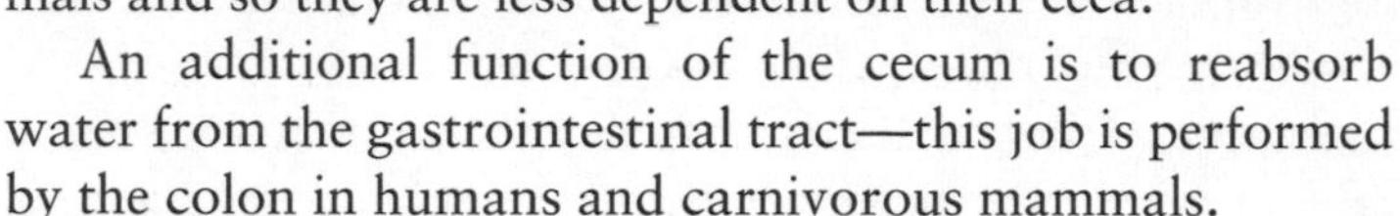

An additional function of the cecum is to reabsorb water from the gastrointestinal tract—this job is performed by the colon in humans and carnivorous mammals.

Richard Luong

Your surgeon was a little out of date. Although it used to be believed that the appendix had no function and was an evolutionary relic, this is no longer thought to be true. Its greatest importance is the immunological function it provides in the developing embryo, but it continues to function even in the adult, although it's not so important and we can live without it.

The function of the appendix appears to be to expose circulating immune cells to antigens from the bacteria and other organisms living in your gut. That helps your immune system to tell friend from foe and stops it from launching damaging attacks on bacteria that happily coexist with you.

There are other parts of the body that appear to do the same thing. Peyer's patches in the intestine help to expose your immune system to the usual contents of the intestine. By the time you are an adult, it seems your immune system has already learned to cope with the foreign substances in the gastrointestinal tract, so your appendix is no longer important. But defects in these immune sampling areas may

be involved in autoimmune diseases and intestine inflammation.

Interestingly, the appendix has been used as a personal "spare part" in surgery. It can be removed and its tissue used in reconstructive surgery of the bladder without risking the immune reaction that would be triggered by using tissue from another individual.

KATHLEEN JAMES

The true appendix is a worm-like, narrow extension beginning abruptly at the apex of the cecum. It is present only in anthropoid apes (gibbons, orangutans, chimpanzees, and gorillas), a few rodents (rabbits and rats), and a few marsupials such as the wombat and the South American opossum. However, in many herbivorous mammals, the large appendix-like pouch of the cecum is an alternative site for fermentation of food. It contains microorganisms that break down cellulose in plant cell walls.

In humans, it was thought to have no physiological function. However, it is now known to play a role in fetal immunity and in young adults. During the early years of development, the appendix functions as a "lymphoid organ," assisting with the maturation of B lymphocytes (a type of white blood cell) and in the production of immunoglobulin A antibodies. In addition, at around the eleventh week of fetal development, endocrine (hormone-producing) cells appear in the appendix. These cells produce peptide hormones that control various biological mechanisms.

JOHAN UYS

Head Weight

How much does a human head weigh? Obviously I can measure the volume of my head by simple water displacement, but I can't tell its density, nor can I work out the weight and density of its various components. Can any of your readers help me out?

BRUCE FIRSTEN

Measuring the weight of your head involves effectively isolating it from the rest of your body. Decapitation has the obvious disadvantage of your not being alive to see the results. However, there is a solution. Your neck vertebrae are responsible for holding your head's weight. If you hang upside down from your feet the vertebrae in your neck move apart slightly because of the weight of your head pulling on them.

To weigh your head you must simply lower yourself slowly onto a scale while hanging upside down. You continually observe the distance between the top vertebra of your neck and your skull, using, say, an ultrasound scanner, and the instant the vertebra starts moving toward the skull you stop and read the scales. Because your neck is not imparting any force on to your head, this isolates your head from your neck, thus giving an accurate measure of your head's weight.

ANDY PHELPS

As a canoeist and kayaker, I remember that when I was learning to do an Eskimo roll my instructor told me to make sure that however much I needed a breath, the last thing to leave the water as my body emerged should be my head. He said the average human head weighs around 10 pounds. Unfortunately, I found that to be a lot of extra weight to lift clear of the water using only the blade of a paddle!

ANDY WELLS

Andy Wells seems to be about right in his recollection of a head's weight. We weren't able to measure the weight of a head directly but we did measure its volume and guess its density on the assumption that the brain, like the rest of the body, is mostly water and we know the density of water at 32°F.

To measure the volume of the head a suitable, virtually bald, volunteer from New Scientist *lowered his head into a bucket of water filled to the brim. The water was as near to 32°F as the volunteer could bear and his head was lowered vertically and crown downward until the water reached the base of the chin. The water that spilled over the sides of the bucket collected in a larger bowl in which the bucket was standing, and its volume was measured. This was repeated five times. The average volume of water displaced was 260 cubic inches, giving an estimate of the weight of a human head at 9.4 pounds.—Ed.*

Raising an Army

After a friend complained about the overzealous attentions of a lover, I came across a reference in Paul Ferris's Sex and the British *to the use of bromide in tea as a means of curbing soldiers' sexual appetites. Is this advice I could pass on to my friend? If so, where would I purchase bromide and what dose is recommended?*

CHLOE DEAR

In the nineteenth century, bromide salts were used widely as sedatives to treat everything from epilepsy to sleeplessness. The bromide salts were said to "reduce the excitability of the brain." The normal dose was between 5 and 30 grains and was taken several times a day (there are about

437 grains per ounce). In the nineteenth century it was not uncommon for children of the upper classes to be flattered by the gift of a personal salt cellar for use at meals. They were led to believe this indicated their increasing status within the family group. The salt was in fact mixed with bromide to make the child better behaved.

MARK WAREING

Bromides are used as a sedative. The libido reduction is really a side effect. The use of bromide salts as a sleeping aid appears in the novels of Émile Zola, indicating their effects were recognized at some time in the nineteenth century. In a reference to using bromides to reduce libido, the comic and author Spike Milligan wrote in *Rommel? Gunner Who?*: "I don't think the bromide had any lasting effect, the only way to stop a British soldier feeling randy is to load bromide into a 300-pound shell and fire it at him from the waist down."

JOHN ROWLAND

In the 1950s, my service in the Air Force as a national service medic, responsible for pharmacy duties and inspections of mess facilities, led me to conclude that the idea of servicemen's tea being laced with bromide was a myth. Nevertheless, there was a strong and widespread belief among recruits during initial training that the apparent loss of male libido must be due to bromide in their tea.

CLIVE HARRIS

I joined the army in late 1945 and remember the suspicion that our tea was laced with bromide. It did taste awful, but most of us, except the very gullible, assumed the story was an "old soldiers' tale" intended to alarm new recruits. The real reason for our lack of libido was exhaustion brought on by physical training. All we wanted to do was sleep.

DAVID ELLIOT

Does my Butt. . . ?

I recently remarked to a female friend of mine that a lot of girls wear black trousers and denim jackets. She told me it was because black trousers "make your butt look smaller." Is this true? Can it be scientifically proved?

NEIL TAYLOR

Yes, your butt does look smaller when you dress in black, at least if viewed from behind.

The reason is that we can perceive shapes only if what we see appears in different shades or colors. If you wore white trousers the shape of your behind could be inferred from the slight shadows cast by its contour. In black clothing, the shadows are invisible and the shape appears flat.

This is the reason why people with dark skin often seem to age well compared with pale-skinned people. Wrinkles and lines, which are visible mainly by virtue of the fact that they create shadows, are harder to detect on darker skin. It is also the reason why facial features need to be greatly exaggerated on dark bronze sculptures.

Of course, your bottom will reveal its true size in profile, but black, especially matte, will save you a lot of exercise and dieting.

GLYN HUGHES
Industrial designer and sculptor

The claim is true and it is because of the uniform dark color of the clothing. Our perception of the shapes of surfaces, among other things, depends on minute shadows and patterns on the surface. See how much easier it is to notice wrinkles on a light shirt than on a dark shirt.

Patterns also play a part. For example lines that start off parallel, then move away from each other, and then return to parallel in a fish-eye effect give the perception of a bump on a smooth surface even if one is not present. In the case of uniform dark trousers, it is very hard to make out the shadows and any pattern, and that's why it's difficult to perceive the exact size of the bump.

LAKSHMI CHAKRAPANI

This isn't a scientific answer but it is evidence in favor. In *Star Trek*, some female cast members supposedly wear padding to increase their female allure. They have two different sizes of padding, and the larger of the two is used when they are wearing dark clothes.

ROB IVES

A similar illusion occurs with striped clothing, but it depends on whether the stripes are horizontal or vertical. Horizontal stripes give the illusion of widening while vertical stripes make a person taller and more compact. So a larger person wearing vertical stripes will take on a more flattering shape. But if people wear horizontal stripes, their proportions will be accentuated . . . sideways.

COLIN VASEY

Mr. Blobby

This is a bit of a girly question, but what exactly is cellulite? There's loads of information on the Web about creams or remedies that promise to miraculously make it disappear, but not much about what it really is.

CATHY TURNER

Such miraculous claims are characteristic of quackery, one of the biggest parasitic industries on the planet, ranking with politics, recreational drugs, and bad-faith litigation. The word "cellulite" was coined to exploit rich innocents.

It has no clear definition, and that is why quacks say so little about what it is.

Cellulite amounts to subcutaneous fat tissue that has accumulated to the point where it bulges between strands of connective tissue, forming an untidy grid like mud oozing through a duckboard. It is found mainly in people who are beyond their first youth, and getting rid of cellulite is just getting rid of fat.

What makes the subject so notorious is that cellulite forms most offensively where the body is least inclined to consume fat deposits, so reducing it takes persistent good dietary sense. No fancy exercise machine, flashy cream, or black box changes that.

For sound information about a large range of such subjects, www.quackwatch.org is an invaluable public service. Don't miss it. Use this website along with http://urban legends.about.com and you will have an antidote to scare stories and a defense against medical fads and scams. Quacks and faddists hate, hate, hate it, and you may read their educational critiques on its "Cheers and Jeers" page.

JON RICHFIELD

Cellulite, the lumpy substance that resembles cottage cheese and is commonly found on the thighs, stomach, and bottom, is just a fancy name for deposits of fat that push against the connective tissue beneath your skin. These cause the surface of the skin to look rather like the dimpled or puckered surface of orange peel. You can check to see if you have cellulite by pinching the skin around your upper thigh. If it looks a bit lumpy, you probably have it. And if you do have it, you're definitely not alone. Most girls and women, and some boys and men, have cellulite.

The amount of cellulite a person has is influenced by several factors. Your genes, your gender, the amount of fat you carry, your age, and the thickness of your skin all affect how much cellulite you have and how visible it is.

Whatever the causes, it's important to know that there

aren't any miracle products, treatments, or medicines that can eliminate cellulite. For example, some fancy salon treatments that promise to get rid of cellulite through deep massage simply reduce the appearance of cellulite temporarily by causing your skin to puff up. Treatments like liposuction and mesotherapy (the injection of drugs to destroy cellulite) either are expensive or, at best, produce only temporary improvement.

To reduce the amount of cellulite you have, the best thing to do is to get rid of excess body fat by eating fewer calories and less fat, as well as exercising more. Experts agree that an exercise routine that combines aerobic exercise with strength training is your best weapon. If you want to conceal your cellulite in the meantime, try using a self-tanner, because cellulite tends to be slightly less noticeable on darker skin.

CATRIONA MACGILLIVRAY

Skin Creep

My three-year-old son stumped me this morning. He was wiping something from the window with his bare hands and wanted to know why his skin squeaked on the glass. I didn't know. Does anybody else?

DAWN HANNA

There are many situations in which a wiping or rubbing action becomes converted to a high-frequency vibration, and the usual reason is the stick-slip nature of friction.

When you start to push one surface over another, friction resists the movement. But if the pushing force is increased, it eventually reaches the threshold at which friction is overcome and slipping starts. At this stage the force required to move the objects drops, and the two surfaces skid over each other.

If one of the objects is elastic, as skin is, it will respond to the increasing force by distorting. When a fingertip is stuck on glass by friction, the skin is initially pulled out of shape by the force attempting to move it, and then springs back closer to its normal shape once sliding starts. But because of this change of shape, the effect of friction once more increases, and the movement of the finger momentarily stops while the skin is pulled into a distorted shape once more. A steady pull will create hundreds of distortions per second, setting up audible sound waves.

But why does friction behave like this? All surfaces are rough on a microscopic scale, and when they come into contact only the highest points, known as asperities, actually touch. These asperities tend to interlock and oppose any movement. As you push harder, the area of true contact increases, because any soft surface, such as skin, deforms to fit more closely into the irregularities of the other surface. The frictional force opposing motion therefore increases. As soon as the finger starts to slide, the asperities bounce off each other and interfere less with the movement.

RICHARD HANN

The skin does not make a noise. The glass, on the other hand, does. Any material, from paper to titanium, when suitably stimulated, is capable of vibrating. Glass is no exception. In a pane of glass there are, as in any other material, definite harmonic series, and the type of rubbing will determine the "note" or harmonic generated.

Not only fingers cause glass to squeak. When being cleaned with detergent and newspaper, the glass will squeak louder and with finer pitch control. If you move the paper slowly, you will hear a moan that, on a large enough piece of glass, sets the whole pane vibrating. Move it more quickly over a small area, and you will produce a tone guaranteed to set a mouse's teeth on edge.

MARTIN JAMES

What Goes In . . .

Is there a formula for working out how much excrement is produced from a certain amount of food? For example, if you ate 1 pound of food, how much excrement would it produce? And how much feces will the average adult human produce each day and what is its composition?

NIGEL WATKINS
and
DAVID BAXTER

One of the major functions of the colon is to absorb water and produce plasticine-like feces which can be voided readily and at will. Feces consist of 75 percent water; bacteria make up half the dry weight; and the rest is unfermented fiber and excreted biliary compounds.

The range of fecal weights produced by individuals varies between 0.75 ounce and 0.66 pounds per day, although this is higher if you have diarrhea and varies between individuals over time. Stool weights in Africa and Asia are said to be double these figures. The only way to increase fecal weight is to eat more fiber. This is because unfermented fiber can hold a lot of water.

Less importantly, some fibers that are fermented in the colon may increase the growth of microbes. And pectin or gum arabic, for example, also yield hydrogen, methane, and short-chain fatty acids. The production of short-chain fatty acids has a possible beneficial action on the bowel mucosa. Products of bacterial fermentation may have an osmotic effect on fecal mass.

Wheat bran is minimally fermented and efficiently increases fecal weight. The coarser and less processed the bran, the more water it can hold and the larger the effect. Whole wheat bread has little or no effect on fecal weight. The increment in fecal weight per gram of wheat bran varies.

In healthy people, the wet fecal weight is on the order of

3 to 5 ounces per ounce of fiber. In individuals with irritable bowel syndrome and symptomatic diverticulitis it is nearer 2 ounces per ounce of fiber.

The effect of the fiber in the colon may be summarized as feces weight = Wf(1 + Hf) + Wb(1 + Hb) + Wm(1 + Hm) where Wf, Wb, and Wm are respectively the dry weights of fiber remaining after fermentation in the colon, bacteria present in the feces, and osmotically active metabolites and other substances in the colon which could reduce the amount of free water absorbed; and Hf, Hb, and Hm denote their respective water-holding capacities.

MARTIN EASTWOOD

People produce up to half a pound of excrement or feces each day. It consists of 75 percent water and 25 percent solid matter. The solid matter is made up of a number of indigestible materials such as fruit skins (33 percent); dead bacteria, which normally live in the gut (50 percent); inorganic matter such as calcium salts; cells shed from the gut; intestinal secretions, including mucus; and bile pigments, which give it its color.

How much excrement you actually produce depends not just on how much food you have eaten, but also on the type of food and the activity of the bowel. If you eat lots of high-fiber foods such as vegetables, beans, and cereals, which the body can't completely digest and absorb, you will produce more feces than if you eat lots of easily digested low-fiber foods, such as chocolate.

Spicy foods, drugs such as laxatives, and infections can affect the activity of the bowel. The greater the speed of transit, the less water the gut can absorb and the greater the weight of feces produced.

JENNIFER KELLY

Natal Knots

Do midwives actually tie a knot in the umbilical cord after birth? If not, what surgical procedure do they perform? And what happened in the days before modern medicine standardized the process?

JACK WHYATT

Umbilical cords consist of three blood vessels and a surrounding medium called Wharton's jelly, the whole lot enclosed by a sheath. This makes the cord too thick to be tied in a knot. Today, where medical supplies are readily available, a plastic clamp is used to compress the cord and cut off the blood supply. Then the cord is cut just above the clamp with scissors.

A clean piece of string or anything that can be tied tightly around the cord, such as a strip of leather or strong grass, would serve just as well if clamps aren't available. A knife, a piece of sharp flint, or even sharp and determined teeth would do instead of the scissors.

The clamp is taken off the cord after three days, and the cord itself rots through a process of dry gangrene and falls off between five and ten days after birth.

SARAH CARTER

To my knowledge, tying an actual knot in the umbilicus after birth has never been practiced, certainly not in my experience as a midwifery nurse.

These days, at delivery, the midwife places two artery forceps on the umbilical cord. She makes a cut between the two forceps using sterile surgical scissors, to separate the placenta. A plastic cord clip is then attached to the cord on the baby's side of the remaining forceps, around 1 inch from the belly button. The final piece of umbilical cord shrivels up and falls off within a few days.

In the past, a couple of different methods have been used.

In the 1960s, during my training, midwives adopted the same procedure but used a sterile rubber band instead of the clip. Before this, midwives would tie the umbilical cord with lengths of string.

MARY COLE
Midwife

When our daughter was born nearly nine years ago, her umbilical cord was sealed off with a small plastic clamp. After a few days the cord shriveled up and dropped off of its own accord. We then found that the clamp was ideal for holding our muesli bag closed. It lasted for another few years until it eventually broke and we were forced to have another child. His clamp is still going strong.

ROB IVES

Thunk!

Does beheading hurt? And, if so, for how long is the severed head aware of its plight?

WILLIAM WILD

Yes, beheading hurts. How much depends on the executioner's skill, or lack of it.

When Mary, Queen of Scots, was executed at Fotheringay Castle in 1587, a clumsy headsman gave her three strokes without quite managing to sever her head. The headsman then had to saw though the skin and gristle with his sheath knife before the job could be regarded as complete. The profound, protracted groan Mary gave when the ax first hit left the horrified witnesses in no doubt that her pain was excruciating.

How long is the interval of consciousness after the head is severed? In France, in the days of the guillotine, some of the condemned were asked to blink their eyes if they were still conscious after the knife fell. Reportedly, their heads blinked for up to 30 seconds after decapitation. How much of this was voluntary and how much due to reflex nerve action is speculation. Most nations with science sophisticated enough to determine this question have long since abandoned decapitation as a legal tool.

DALE MCINTYRE
University of Cambridge, UK

Antoine Lavoisier, the French chemist who lived between 1743 and 1794, was caught up in the revolution and faced beheading. He asked friends to observe closely as he would continue blinking as long as possible after being killed. He was reported to have blinked for 15 seconds after decapitation.

A. GRYANT

The story of Antoine Lavoisier's last heroic service for science has been reported many times but unfortunately appears to have no basis in fact. It is not given in any contemporary account we have been able to find, or in the standard accounts of his life and death. As pointed out above, however, there have been attempts to ascertain if a severed head retains consciousness. The most reliable account appears to be that given below.—Ed.

A particularly detailed report comes from Dr. Beaurieux, who, under perfect circumstances, experimented with the head of the murderer Languille, guillotined at 5:30 A.M. on June 28, 1905. (From *A History of the Guillotine* by Alister Kershaw. His source is *Archives d'Anthropologie Criminelle*, 1905.)

> Here, then, is what I was able to note immediately after the decapitation: the eyelids and lips of the guillotined

man worked in irregularly rhythmic contractions for about five or six seconds. . . . I waited for several seconds. The spasmodic movements ceased. The face relaxed, the lids half closed on the eyeballs, leaving only the white of the conjunctiva visible, exactly as in the dying whom we have occasion to see every day in the exercise of our profession, or as in those just dead. It was then that I called in a strong, sharp voice: "Languille!" I saw the eyelids slowly lift up, without any spasmodic contractions. . . . Next Languille's eyes very definitely fixed themselves on mine and the pupils focused themselves. . . . After several seconds, the eyelids closed again, slowly and evenly, and the head took on the same appearance as it had had before I called out.

It was at that point that I called out again and, once more, without any spasm, slowly, the eyelids lifted and undeniably living eyes fixed themselves on mine with perhaps even more penetration than the first time. Then there was a further closing of the eyelids, but now less complete. I attempted the effect of a third call; there was no further movement and the eyes took on the glazed look which they have in the dead.

I have just recounted to you with rigorous exactness what I was able to observe. The whole thing had lasted twenty-five to thirty seconds.

For further details see www.metaphor.dk/guillotine

MIKE SNOWDEN

If indeed a severed head remains conscious for a short while, then the following procedure might be regarded as humane, assuming the purpose was to convince the dying man he was flying to heaven.—Ed.

Dr. Livingstone wrote that Africans he encountered were aware that consciousness is not lost immediately. He recounts how they bent a springy sapling and tied cords from it under the ears of a man to be decapitated so that his last few moments of awareness would be of flying through the air.

JOHN RUDGE

However quickly consciousness is lost, there is little doubt that the procedure must produce a painful few seconds. In 1983, Harold Hillman, then reader in physiology at the University of Surrey, wrote an account of the suffering caused by different methods of execution for *New Scientist* at the time when the World Medical Association had just discussed attitudes of physicians to capital punishment. This is what Hillman said about the guillotine:

> The guillotine was named after the French deputy who proposed the use of the device in 1789. It was tested on corpses at the Bicêtre Hospital in Paris, and employed by the French Revolution in 1792. It was introduced as a swift and painless device, as Joseph-Ignace Guillotin believed, to extend to all citizens the advantages of a technique used only on noblemen. Although people believe that Guillotin invented the device, it had been used in Italy, Germany, France, and Scotland in the sixteenth century.
>
> Guillotining was considered more humane because the blade was sharper and execution was more rapid than that accomplished with an ax. Death occurs due to separation of the brain and spinal cord, after transection of the surrounding tissues. This must cause acute and possibly severe pain. Consciousness is probably lost within 2 to 3 seconds, owing to a rapid fall of intracranial perfusion of blood.
>
> There are accounts of the eyes looking around from the severed head, and animals may do this when they are guillotined for experiments in which their organs are to be excised or their brain biochemistry is to be examined rapidly.

Thanks to Tony Corless for drawing our attention to this article.

Bodily Breeding

How many different species live on or in the human body and what is the actual total population of these guests?

ROGER TAYLOR

The microorganisms that inhabit the body of a healthy human being are known as the normal microbial fauna and they come in two different types—those that are permanently resident and those that are transient. Of course, any number of fascinating and nasty parasites can join this microbial community and make the human body their home.

In his seminal work Life on Man *(Secker & Warburg, 1969), the bacteriologist Theodor Rosebury gives a full biological and historical account of the microbes that live on the average human. The numbers involved are huge, Rosebury tells us: "If we are to get to the microscopic center of this with our eyes open and our stomachs steady we might do better to look gingerly and sip instead of gulping. . . . The life on man consists of microbes in extraordinary variety and large numbers."*

The figures that he grapples with are quite mind-boggling. For example, he counted 80 distinguishable species living in the mouth alone and estimated that the total number of bacteria excreted each day by an adult ranges from 100 billion to 100 trillion. From this figure it can be estimated that the microbial density on a square inch of human bowel is around 50 billion organisms.

Microbes inhabit every surface of a healthy adult human that is exposed to the outside, such as the skin, or that is accessible from the outside—the intestines, from mouth to anus, plus eyes, ears, and airways. Rosebury estimates that 50 million individual bacteria live on the average square inch of human skin, describing the surface of the body as akin to a "teeming human population during Christmas shopping."

However, this figure can vary widely throughout the almost 20 square feet that make up the surface area of a human. In the oily skin that is found on the side of the nose or in a sweaty armpit, the figure can increase tenfold; and once inside the body, on the surface of the teeth, throat, or alimentary tract, these concentrations can increase a thousandfold. These inside surfaces are the most densely populated region of the human body.

Conversely, on those surfaces where there is liquid flow removing bacteria, such as the tear duct or genitourinary surfaces, the populations of organisms are much thinner. Indeed, Rosebury could detect no microbial life at all in the bladder and lower reaches of the lungs.

Yet, while the figures appear huge, he estimates that all the bacteria living on the external surface of a human would fit into a medium-sized pea; and all those on the inside would fill a vessel with a capacity of a mere 10 ounces.

If disease organisms such as viruses or other infections are present, these figures increase, but not by any significant amount. The total number of organisms living on us is huge, but when one considers the volume of the human body, the volume of species using us as home is not so great.

As to the total number of species that are inhabiting a healthy body, estimates vary as more species are discovered on a seemingly regular basis, but Mark

Pallen, a professor of microbiology based at the Queen's University of Belfast, reckons that the figure is in excess of 200.

"There are more than 80 that live in the mouth alone and studies that have been carried out at the Laboratory of the Ecology and Physiology of the Digestive System in Jouy-en-Josas, France, suggest that at least another 80 live in the gut, with many others living on our skin. It's impossible to be precise, but our permanent resident population certainly exceeds 200 species," Pallen says. "The human genome carries a maximum of 100,000 genes, yet the average bacterial genome has 2,000 genes. Therefore there are actually four times as many genes found in the bacteria that live on humans as there are in the human genome itself."

Of course, it's not just bacteria and viruses that make people their home. In his books Fearsome Fauna *(W. H. Freeman, 1999) and* Furtive Fauna *(Penguin, 1992), Roger M. Knutson describes the wide range of parasites that live both on and inside you. These tend to be macroscopic organisms, and some of them can be pretty gruesome creatures.*

Lice are perhaps the most common of these body dwellers. They have the ability to get everywhere from your hair to your armpits to your groin. Nonetheless, they tend to be more itchy than damaging, unlike ticks, which can cause any number of nasty and exotic diseases from royal farm virus to Omsk hemorrhagic fever. And then there is the scabies mite, which is believed to infest millions of humans worldwide, and is able to burrow into the body to hide itself, causing a nasty itch.

Fortunately, its close relative, the follicle mite, which is found on everybody in the world, happily munches dried skin cells and causes far less irritation. And not all body parasites creep and crawl—

you can find fungi in your hair and mold in your skin folds if you look closely enough.

Inside your digestive tract you can, among others, find the protozoan that causes amebic dysentery, 60-foot-long beef tapeworms, and a hookworm that has a penchant for finding its way into your bloodstream.

Other creatures in your blood can include the hermaphroditic Shistosoma worm, which can lead to a bloody and scarred bladder, while in your lymphatic system you may find the five-inch Wucheria worm. In your liver you may come across the bile-loving Clonorchis sinensis *fluke; and, perhaps most horrifying of all, the brain can house* Naegleria fowleri, *an amoeba that just loves the warmth that it finds inside your skull, reproducing in its millions until you drop down dead.—Ed.*

Goggle-Eyed

It has always intrigued me that I can see clearly underwater if I wear goggles or a mask, yet if I don't everything is blurred. What is it about my eyes or the water that causes this effect?

MICHAEL SLATER

The reason that you see this effect is the same reason a spoon looks bent when you immerse it in a glass of water. Light travels more slowly through water than it does through air. When light moves from one medium to another, it changes speed, and as a result the beam is bent, or refracted. The amount of bending depends on the ratio of the speed of light through each medium.

The human eye is delicately balanced to make sure that

light coming in through the pupil is focused onto the retina at the back of the eye. But it's optimized for light coming from air and hitting the surface of the eye. The eye has evolved to take account of the refraction that takes place at the interface between air and eye, and gives a focused image on the retina.

However, when light comes directly from water to the eye, the light is bent by a different amount, so the light isn't correctly focused. Goggles restore the air/eye interface and normal sight is resumed.

This phenomenon of light bending when it goes through different media is used to our advantage in spectacles, where lenses bend the light to correct imperfect vision.

RICHARD WILLIAMS

The amount of refraction or bending of the light depends on the difference of the refractive indices of the media on the two sides of the surface of the cornea—the bigger the difference, the more the light is bent. Because the refractive indices of air, water, and the cornea are 1, 1.33, and 1.38 respectively, this difference is much smaller when the eye is in contact with water than when it is in contact with air.

The power P of a surface is given by $P = (n_1\ n_2)/R$, where n_1 and n_2 are the refractive indices of the cornea and the medium outside it respectively and R is the radius of curvature of the cornea. P is measured in diopters. A diopter is the unit of refractive power, equal to the reciprocal of the focal length (in meters) of a given lens. Assuming a value of 0.008 meter for R, the power in air is about 47 diopters and the power in water is about 6 diopters.

The focusing ability of the eye is variable to a degree because the shape of the lens can be controlled by the ciliary muscles.

However, the increase of power which this can produce is much less than the loss of power of 41 diopters when the eye is in contact with water. In fact, the biggest change the eye can achieve is about 15 diopters in young children, dropping to about only 1 diopter in 60-year-old adults. This means that the eye in water is unable to bend the light entering it enough to focus it onto the retina, so objects appear blurred.

William Madil

2 Plants and Animals

Chorus Line

Ever since finding a millipede in my bath, I've wondered why this creature has so many legs. What advantage do they provide and how did it get them?

SARAH CREW

Millipedes and earthworms have similar lifestyles. Both burrow in soil, eating dead and decaying vegetation, but they have evolved very different methods for forcing their way through the soil. Worms use the strong muscles in their body walls to build up pressure in the body cavity, and so develop the forces needed to push forward or widen a crevice in the soil. Millipedes, however, use their legs to push through the soil. The more legs the animal has, the harder it can push.

Millipedes are different from centipedes. They have very large numbers of short legs because long legs would be a liability in a burrow. Centipedes, which spend their time on the surface or among leaf litter, have fewer, longer legs. They have little need to push, but have to run faster than millipedes.

Millipedes, centipedes, and earthworms all have long, slender bodies, divided into large numbers of segments. Except at the two ends of the body, all the segments are built to more or less the same design. Similarly, many products of human engineering are built largely from a series of identical modules. For example, identical seats and windows are repeated many times along the length of a bus. The advan-

tage of this is that one design will serve for all the seats, and one machine can make them all.

In the same way, the repetition of segments in animals reduces the quantity of genetic information needed for development. Millipedes presumably evolved from an ancestor with fewer segments and correspondingly fewer legs, simply by changes in the genes that specify the number of segments.

R. McNeill Alexander
Emeritus Professor of Zoology
University of Leeds, UK

Fly, Fly Away

While we were on vacation, flies were a daily nuisance when we were sitting on the hotel balcony, yet one evening the wind picked up and the flies disappeared. It seems there is a wind speed above which flies are unhappy. What is it, and why? Where do the flies go and do these variables differ between species?

Bill Williamson

There is indeed a maximum wind speed above which they will not fly, but it also depends on temperature, humidity, and the sex and age of the fly, as well as its species.

For *Lucilia cuprina*, the major sheep-infesting fly in southern Australia, the wind speed threshold is about 20 miles per hour. This is the upper limit at which these flies can maneuver safely away from objects that would damage their wings. However, some fly species depend on wind transport to carry them long distances. The annoying tiny Australian bush fly is believed to migrate to Tasmania this way each summer.

Lucilia cuprina is also grounded if it is hotter than about 104°F, and below about 54°F. When conditions stop them from flying, they crawl into a protected space and wait until things improve.

Many species of fly live relatively short lives, measured in days. The females lay eggs or living young, according to species, which first live as a maggot and pupate underground for up to a year or more. They emerge only when conditions are right—fine, clear, sunny days with little wind, perfect to annoy the tourists.

JAN HORTON

A cyclist friend informs me that, after extensive experiments, members of Melbourne Cycling Club have concluded that they cease to be troubled by flies when riding at or above 10 miles per hour in still air.

TONY HEYES

I live in subarctic Canada, where millions of mosquitoes appear during our short summer. When it is windy they hide in vegetation close to the ground, where they won't get blown away. Walking across an apparently empty tundra will raise thousands, anxious for a quick meal of your blood.

MICHAEL MORSE

I can confirm that the maximum speed of certain flies is about 10 miles per hour. When I was cycling below that speed up mountains in the Alps and Pyrenees, flies were a constant nuisance. Not having the fitness of Lance Armstrong and laden with luggage, I found that speed difficult to sustain while climbing. It was a choice between exhaustion and grisly insect attack—I chose the latter. When the flies are outpaced, I can only presume they land on the next cyclist struggling up the same road.

STEVE LOCKWOOD

Living Bath

I have made a birdbath in my garden with a large plastic receptacle fashioned from a plant-pot saucer to hold the water. It has proved to be successful, with many birds using it each day, but I have found that algae build up very quickly on the surface covered by the water. The plastic receptacle was bought new from the shop and the water I put in it is good drinking water, so where do the algae come from? The birds bathe in it and drink the water without any obvious ill effects.

CALVERLEY REDFEARN

There are two issues here. Where do the algae come from, and how do they survive and grow?

Freshwater algae are well adapted to distribute and establish themselves in new, often temporary, ponds and puddles. Most green alga species can make tough-walled resting spores when conditions get difficult, usually when their habitat dries up. These dry resting spores can survive a long time and are small enough to be picked up and carried by the wind or in mud that has dried onto the feet of birds. Any newly filled birdbath will very rapidly be colonized, with the spores revitalizing as soon as they enter water.

But the algae also need nutrients to grow, and wouldn't survive well in distilled water. Ordinary drinkable tap water, processed to be free of toxins and microorganisms, will still contain quite high concentrations of plant nutrients such as nitrates and phosphates. Other sources include dust carried in the air and, of course, general muck on the bodies of the birds bathing in the bath.

For these reasons, repeatedly topping up a garden pond with tap water usually leads to the unsightly green algal overgrowth your correspondent has reported.

So it is absolutely normal for algae to grow in birdbaths,

which are just another pond habitat for them. The good news is that the algae are harmless and won't worry the birds.

STEPHEN HEAD
Director of Pond Conservation
Water Habitats Trust, Oxford Brookes University, UK

Given light and water containing dissolved nutrients, a microscopic amount of algae or their spores will grow to a visible film before too long.

But where does the microscopic amount of algae come from to start with? It might have been on the receptacle, which was not sterilized, or in the water, which also does not have to be sterile to be drinkable. Or it could have come from the surrounding environment, either airborne or carried in by birds.

Those who work in offices that have watercoolers equipped with large, clear plastic bottles will know that drinking water in a plastic container closed to the environment can grow algae if it is left in sunlight for too long. In fact, I once drank a glass of such water before I noticed. Although it tasted slightly different, I subsequently suffered no obvious ill effects.

IAN WILLIAMSON

Siren Screams

When emergency sirens pass by, all the dogs in my neighborhood yowl. The reason, I've read, is that the sound of the siren hurts their sensitive ears. Yet my cat, whose hearing seems to be more sensitive than my dog's, pays no attention. Why would the sound hurt a dog's ears and not those of a cat?

MICHAEL HAM

The reason dogs yowl when emergency services go by may be that to the dogs, the siren sounds like other dogs howling and they respond by howling back. This goes back to the time when they hunted in packs and signaled to one another when searching for prey. Even if the screaming siren does not mimic exactly the sound of another dog, they can probably pick out a component part of the siren that does. Cats, on the other hand, hunt alone, are not pack animals, and so do not respond to the sirens.

ANNE BLOOMBERG

Your correspondent might like to read the excellent *Dogwatching: Why Dogs Bark and Other Canine Mysteries Explained*. In it, the anthropologist Desmond Morris answers 46 FAQs. He mentions that families who attempt to sing music together are sometimes helped, or hindered, by their dog, which joins in when its human family breaks into a group howl.

Dogs, wolves, and humans evolved as cooperative hunters, and more recently, sheep guardians, with a need to keep in touch with their partners on the next ridge. Hence howling, yodelling, and such devices as the Israeli *challil*, or shepherd's flute. Sirens are artificial, amplified howling. Their rise and fall is calculated to alarm and stir us; and to my ear, and my dogs' ears, they succeed splendidly.

ANN BRADFORD DRUMMOND

Shell Shock

A few weeks ago I was in Belgium with friends. Some of the party ate snails in garlic, and one took an empty shell home for his 3-year-old son to play with. The washed shell sat on the kitchen work surface most of the time, until one day two baby snails emerged from it. The "parent snail" had long since been fried, scooped out, and eaten. Assuming my friend is not hoaxing us, what happened here?

Dave Mitchell

Several readers thought this question was indeed an elaborate hoax. But there may be a simple answer. —Ed.

Snails both fertilize and carry their eggs internally. When being prepared for the table, the snails are scooped out of their shells; usually mixed with butter, parsley, and garlic; then cooked. After cooking they are reinserted into their shells and served.

The shells themselves are not cooked, so the baby snails that later emerged had presumably originated from eggs lodged inside the shell. These could have survived the scooping out during the preparation.

Gregory Sams

The snail that is most frequently eaten throughout Europe is *Helix pomatia*. It is known in the United Kingdom as the "Roman snail" because the Romans may have introduced it to these shores for food. It is native to much of Europe.

Snails that are eaten in restaurants often originate from snail farms, although they may be collected from the wild. They are hermaphrodites, but although they have both male and female reproductive organs they must mate with another snail before they lay eggs. After mating, a snail can store the sperm it received for up to a year before fertilization, but eggs are usually laid within a few weeks of mating.

The eggs of *H. pomatia* are laid about 2 inches deep in holes dug in the soil. A snail will take up to two days to lay between 30 and 50 eggs. After about four weeks the fully formed baby snails hatch.

In this case, any eggs that were present during the cooking process would die. However, it is possible that the shell in which the snail was served was not the one belonging to its fried occupant. Snails are often supplied ready-cooked from producers with a separate supply of shells in which they can be placed for presentation, one of which, in this case, may have held eggs previously deposited in the upper whorls of the empty shell.

More information about snails and their conservation may be found on the Web site of the Conchological Society of Great Britain and Ireland at www.conchsoc.org.

PETER TOPLEY

Living on Stone

There is a brick chimney stack near my home that has a tree growing out of it, and I have seen similar trees on rock faces and cathedral spires. How do these plants survive? Where do their roots go? Young trees in my garden struggle in well-tended, well-fed soil. How can a 3-foot tree survive in what is essentially a brick wall?

JANE STEPHENS

The absence of trees in most chimneys shows that a very favorable site is required. Damaged brickwork can provide a crevice for a seed to lodge in and send down a root, usually during a particularly wet year. The extra wetness permits the roots to go deep enough into the brickwork to survive the next year, so the initial damage needs to be substantial.

A factory chimney has thick, solid brickwork, which protects the interior of the bricks from drying. Rain can enter through cracks at the top, and wind can drive it into the chimney from the side. The swelling root widens and deepens the cracks, making the site even more favorable. This process is helped when water freezes and expands in the cracks in winter, and rain dissolves some of the lime in the mortar.

Numerous herbaceous species, such as willow herbs, produce wind-dispersed seed, which can easily reach a stack top and will germinate and grow briefly before dying in the drier weather. Their decaying roots enhance the water-holding capacity of the brickwork before the tree seed arrives.

The 3-foot tree mentioned is only superficially similar to a 3-foot garden specimen. The chances are that it is much older than a garden tree of the same size, and manages very little growth except in wet years. Also, it will be more branched and have smaller leaves darkened by red pigments, which all plants produce under stress. In extremis it can reduce its rate of water loss by shedding leaves.

Of course it may well be an exotic garden species, not a wild one. Buddleia grows everywhere in these rainy conditions, with plants up to 1 foot on flat wall tops. I have seen a plant up to 6 feet tall where a broken gutter pours copious water over brickwork. The roots penetrated and split the wall, which was built with soft lime mortar.

Ian Hartland

Rocky walls can hold enough water for plants, and there is always enough air to provide carbon dioxide. Often bird droppings, dust, and minerals dissolved from the rock supply enough of the other required nutrients. Indeed some epiphytes such as tillandsia, or "air plants," get practically all their mineral nutrition from dust.

Garden trees are not generally epiphytes, and are not well suited to clinging to life among rocks or on bark. But many a fig, especially strangler fig species, starts life as a bonsai growing from a seed in bird droppings on a wall, cliff face, or tree trunk. There the figs cling to life, sometimes for centuries, until time and chance destroy them or lead a rootlet to good soil.

Jon Richfield

I live in an area where a once-thriving quarrying industry has left vast mountains of slate chippings. Now several organizations are using various methods to re-foliate this barren landscape. One method is to bring in truckloads of topsoil and replant the hillsides with saplings in an attempt to landscape the surroundings. There are numerous problems involved: the slate slopes drain exceptionally well and do not readily retain sufficient water for healthy plant growth; the frequent rain washes away the soil; and the slopes are not stable enough for larger plants to maintain their purchase.

A less futile and less labor-intensive method involves throwing lots of seeds for suitable shrubs and trees onto the slate piles, and knocking in perching posts at regular intervals. Birds eat the scattered seed, then rest on the posts, leaving behind their droppings. These contain some active seeds and also act as a fertilizer. The birds may even import seeds from the surrounding countryside.

When the shrubs begin to grow, they provide even more perches for the birds, and focus the fertilizer where it is most needed. These plants promote the growth of smaller plants such as mosses and grasses by providing shade and dropped leaves. The smaller plants help to retain water and begin to build soil as they break down.

I would suggest that this process is responsible for the survival of trees in seemingly soil-free stone and brick structures. Birds nesting or perching in these safe places provide seeds, and their droppings provide the fertilizer. The nutrients also encourage mosses, which retain moisture for the tree and eventually provide a kind of soil as they die and break down. The restriction to the tree roots would act to "bonsai" these trees, so they do not outgrow their nooks and crannies.

JEREMY WATKINS

Toxic Taters

When I was a child my grandmother told me that I should never eat the green areas of skin on old or damaged potatoes. I've since learned that this contains a toxin similar to that found in deadly nightshade. But how much green potato skin would I have to eat before falling ill, and what would the toxin actually do to me? Do similar problems lurk in species related to potatoes, such as yams or eggplants?

EMILY JANE HORSEMAN

Potatoes are a member of the Solanaceae family of plants, which includes tomatoes, peppers, eggplants, tobacco, and deadly nightshade. They are characterized by their ability to produce toxic alkaloids such as solanine in their leaves, roots, and fruit. Unfortunately, solanine does not dissolve, so it cannot be removed by soaking the potato and it is not destroyed by cooking.

Even in edible varieties, a high concentration of glycoalkaloids is present in the leaves, shoots, and fruit and these should never be eaten. Potato tubers should always be stored where it is cool, dry, and dark, because those exposed to light may develop unacceptable concentrations of glycoalkaloids, indicated by the green color you find in aging potatoes.

People have died from solanine poisoning. A report by Canada's Bureau of Chemical Safety for the World Health Organization concluded there is no safe level of solanine in food, though levels of between 5 and 50 milligrams per pound found in correctly stored potatoes are acceptable.

A lethal dose might be 1 milligram of glycoalkaloids per pound of a person's body mass. Therefore, an adult with a mass of, say, 160 pounds might consume a fatal dose by eating around 5 pounds of potatoes. Even if one were hungry enough to eat a meal this large, the more likely outcome would be gastrointestinal or neurological symptoms and not

death. Besides, if concentrations in the potato had reached hazardous levels this would most likely be detected from the bitter taste or the burning sensation in your throat.

It is sometimes said that if the potato were introduced to Europe today, instead of in the sixteenth century, the European Union would have banned it under the Novel Foods Regulation (EC) 258197. This requires all foods that do not have a history of consumption in the EU before May 1997 to undergo a premarket safety check. The importance of checking was dramatically demonstrated in the United States, where a man almost died from eating the Lenape variety, introduced in 1964 without screening for glycoalkaloids.

Potato genetics are complex, and attempts to produce new varieties from the narrow genetic base of the hybrids introduced from Latin America frequently give rise to plants with high solanine concentrations.

MIKE FOLLOWS

When a potato is exposed to light, its solanine content escalates as a natural protection against being eaten by foraging animals. It is, after all, meant to propagate a new plant rather than be consumed. Solanine gives potatoes a bitter taste and checks the action of the neurotransmitter acetylcholine. This causes dry mouth, thirst, and palpitations. At higher doses it can cause delirium, hallucinations, and paralysis.

The green in a toxic potato is harmless chlorophyll, but it acts as a warning that the potato has an elevated level of solanine. The entire potato should be discarded. The same applies to potatoes that have begun to sprout and to potatoes that show black streaks from late blight. The fatal dose of solanine for an average adult is between 1 and 3 milligrams per pound of body weight, or between 200 and 500 milligrams in total, depending on body weight. Properly stored potatoes contain less than 100 milligrams per pound, so a fatal dose could, arguably, be obtained from as little as 2 pounds if a person had a small body mass.

Solanine is concentrated in the potato skin, so peeling removes between 30 and 90 percent of this toxin. That runs counter to the old saying, "The skin is the best part." In the past, potatoes were stored unwashed in paper sacks and dumped on the bottom shelf or in the darkest place in the vegetable store. The modern practice of washing potatoes and packing them in clear plastic increases solanine risk. On exposure to light at 60°F the solanine content quadruples every 24 hours. At 170°F it can be nine times greater and can reach to 800 milligrams per pound in the skin.

Other nightshades, such as tomatoes, eggplants, and capsicums, also contain solanine in varying quantities, depending on the degree of ripeness and whether they are infected with blight. Nicotine in tobacco, another nightshade, is the same type of glycoalkaloid as solanine, but the high temperature of combustion is believed to modify its toxic impact—eating a cigarette is far more toxic than smoking one.

Craig Sams

Eggplants are related to potatoes and also contain solanine in addition to high levels of histamine and nicotine. There are documented cases of allergy to both the fruit and the pollen, and of histamine reactions in sensitive individuals. Nevertheless, most of the toxic components in edible versions have been bred out and there are no wild varieties on sale.

Yams are not related to potatoes, but like many plants they contain toxic compounds including polyphenols and a tannin-like chemical. They also contain a variety of minor alkaloids, some of which are used in the synthesis of birth control pills and corticosteroids. Some varieties of yam contain the bitter alkaloids dihydrodioscorine and dioscorine. These are water-soluble alkaloids that, on ingestion, produce severe and distressing symptoms that can prove fatal. Yams are usually detoxified by soaking in salt water, hot fresh water, or running streams.

Derek Matthews

Mole Holes

As autumn progresses my garden is again blessed with a trail of molehills linked by networks of tunnels barely below the surface. This prompts a host of questions: How big is the average mole tunnel network? Is the mole constantly developing the network and do areas become redundant? How far does the average mole tunnel in its lifetime? And if moles are fiercely solitary, do individual networks overlap? If not, how do moles find each other to ensure future mole generations?

ALAN ROWE

The depth and extent of the tunnel system of a mole (*Talpa europaea*) will vary considerably depending on a number of factors such as the type of soil and the height of the local water table. Earthworms and other invertebrates that enter the tunnel system are the moles' main source of food, so it is likely that a mole living in a worm-rich meadow will need a less extensive tunnel system than a mole that inhabits a tunnel system in an acidic soil where worm numbers are much lower.

Moles do extend their tunnel systems when necessary and they will abandon those that are no longer needed or productive. Their digging activity increases in the autumn when the colder soil temperatures send earthworms (and their mole hunters) deeper below the surface. In the spring earthworms start to return to the surface layers of the soil and there will be much more mole activity as the moles begin to make new surface tunnels or repair old ones.

Moles are largely solitary animals outside their spring breeding season and they will drive out those of their species that intrude into their tunnel systems. However, in areas where mole populations occur at high densities their tunnels may overlap.

During the mating season in February and March, the

males become far more mobile and will frequently leave their territories in search of mates. Much of this traveling is done at ground level but they can also make use of existing tunnel systems. Females are probably located by scent, but very little is known about the mating behavior of moles.

ANDREW HALSTEAD
Principal entomologist
Royal Horticultural Society, London, UK

Most of the tunnels that a mole constructs are actually sophisticated traps for the numerous invertebrates on which it feeds. The tunnels that are designed as traps need to be extensive for moles to catch sufficient food, and their length will vary with the density of suitable edible invertebrates in the soil. In those areas where such invertebrates are few and far between, the tunnels, by necessity, need to be longer and the system more extensive.

Moles are solitary creatures. Groups of them will, however, cooperate to some degree. In places where there is ample food but no water a long tunnel used occasionally by all the mole group will link several groups of traps to a water supply. A pair of moles will also cooperate on land that is subject to periodic flooding to build a heap of earth of sufficient size to support a chamber in which young can be suckled above the water level. Moles are very good swimmers, though not as good as the closely related species, the Pyrenean desman (*Desmana pyrenaica*).

MIKE EASTHAM

The territory of an adult mole usually covers between 20,000 and 70,000 square feet, with males likely to have larger territories than females. Depending on the soil type, there may be as many as six levels of tunnel lying beneath the turf.

Shallow tunnels are created by the mole's pushing its way through the earth and bracing its body to compress the soil into walls around it. However, deeper tunnels require true excavation with the mole digging out soil into the tun-

nel directly behind it, and then performing a somersault and subsequently bulldozing the loose soil up to the surface to form the kind of molehill with which we are all familiar and which is a constant bane of the landscape gardener.

Building an extensive tunnel network requires a substantial investment of labor, which perhaps explains why, once built, the system is fiercely defended. Maintenance of an established tunnel system requires much less effort. Territories do often overlap and, where any tunnels meet, moles leave scent signals to clearly establish their boundaries. If an owner is absent for any length of time and those scent markings disappear, the tunnels will be taken over very quickly by rival moles.

LILLIAN WALKER

Walking Tall

During a recent visit to Kenya I noticed that giraffes tend to walk with a "pacing" gait. The two legs on each side of the animal move at the same time, unlike those of horses and other four-legged animals. As far as I am aware no other ruminant (aside from the camel and okapi) walks in such a fashion. Does anybody know the biomechanical reasoning behind this and indeed whether it is unique to the giraffe and camel? Is this form of locomotion more efficient than a conventional gait?

ROGER SANTER

Giraffes and camels have long legs, relatively short bodies, and large feet. A common explanation for their unusual gaits is that the gait prevents fore and hind feet from getting in each other's way.

If you refer to feet by their initials, LF for left fore, RH for right hind, and so on, you can write down the walking patterns of particular animals. When most mammals walk, they move their feet in turn, always in the same order, at more or less equal intervals of time:

LF**RH**RF**LH**LF**RH**RF**LH

and so on. The asterisks indicate the time intervals, **** for long and * for short.

In trotting, which is a faster gait, the legs move two at a time, in diagonally opposite pairs:

(LF + RH)****(RF + LH)****(LF + RH)****(RF + LH)

and so on.

Camels, however, do something different. Instead of trotting, they pace, moving the two feet on the same side of the body together:

(LH + LF)****(RH + RF)****(LH + LF)****(RH + RF)

and so on.

The questioner says that walking giraffes move the two feet of the same side of the body at once, like the camel, but that isn't quite true. Films analyzed by the American zoologist Milton Hildebrand showed that walking giraffes move their feet like this:

LF***RH*RF***LH*LF***RH*RF***LH

and so on.

Long and short time intervals alternate and the forelegs move slightly after the hind legs of the same side.

In trotting, a foreleg swings back while the hind leg on the same side swings forward, so there is a danger of the feet colliding, if the legs are long. In a pacing walk, the legs on one side of the body both swing forward, then both swing back, so the fore and hind feet are kept well out of each other's way.

The fact that some long-legged breeds of dog also pace instead of trotting supports this explanation of the giraffe's pacing walk.

In horses, there is less danger of fore–hind collisions in the standard walk than in the trot. The giraffe's pace-like walk reduces the danger further. This may explain the giraffe's unusual gait. But I feel bound to point out that both camels and giraffes gallop successfully. In galloping, both forelegs swing back while both hind legs are swinging forward, offering plenty of scope for collisions.

Hildebrand recorded giraffe-like walking gaits for the cheetah, a hyena, and the gerenuk (a long-legged antelope). No experiments seem to have been done to find out whether there is any difference in energy costs between trotting and pacing, or between the horse's and the giraffe's styles of walking, but I would expect any differences to be small.

R. McNeill Alexander
Emeritus Professor of Zoology
University of Leeds, UK

Who Needs Nine Lives?

A friend of mine reckons that you can drop a cat from any height and it will survive unhurt because its terminal velocity is lower than the speed at which it can land unhurt. Can someone confirm or refute this, because kittens in my house now look strangely at my friend. I'm sure this can't be true, can it?

Anna Goodman

I'm reminded of a study reported in the *Journal of the American Veterinary Medicine Association* in 1987 by W. O. Whitney and C. J. Mehlhaff, two New York vets, entitled "High-Rise Syndrome in Cats." The study was also summarized in *Nature* a year later.

Briefly, the authors examined injuries and mortality rates in cats that had been brought to their hospital following falls ranging from between two and 32 stories. Overall mortality rates were low, with 90 percent of the cats surviving, a fact that supports the correspondent's ailurophobic friend. However, the study unexpectedly found that the incidence of injuries and death peaked for falls of around seven stories, and then actually decreased for falls from greater heights.

The *Nature* article presents three main variables that determine injury and mortality rate—the speed reached by the cat, the distance in which said cat is brought to a stop, and the area of cat over which the stopping force is spread. While concrete streets work in nobody's favor when it comes to stopping falling items, cats suffer relatively little injury (compared with their owners) because they do indeed reach lower terminal velocities and absorb the shock of stopping so much better. A falling cat has a higher ratio of surface area to mass than a falling human, and so reaches a terminal velocity of about 65 miles per hour (about half that of humans). Cats are also able to twist themselves so that the impact is spread over four feet, rather than our two. And as they are more flexible than humans, they can land with flexed limbs and dissipate the impact forces through soft tissue.

To answer the paradoxical increase in survival rates once seven stories have been reached, the authors suggested that an accelerating cat tends to stiffen up, reducing its ability to absorb the impact. However, once terminal velocity is reached, there is no longer any net force acting on the cat, and so it will relax, increasing both its flexibility and the cross-sectional area over which the impact is dissipated once the cat hits the ground.

I'd still keep your friend away from your kittens, if I were you.

John Bothwell
Marine Biological Association, Plymouth, Devon, UK

When cats land, they bend their legs to absorb the shock as we do at our knees. Obviously this action will lower their bodies toward the ground, especially their heads, thanks to their having four legs. Above a certain height, the bending action will bring the chin into contact with the ground and so cats dropped or jumping from great heights will contact the ground with force, shattering their jaw.

NIKKI LOUGH

Vets quite commonly see jaw injuries in cats, usually as a result of their taking too high a leap from a wall.—Ed.

I don't know what the terminal velocity of the average cat is, but this question did remind me of a joke.

Because cats always land on their feet and toast always lands buttered side down, you can construct a perpetual motion machine by simply strapping a slice of buttered toast to a cat's back. When the cat is dropped it will remain suspended and revolve indefinitely due to the opposing forces.

CATHERINE

The risks to different animals of taking a fall were laid out in 1927 by the biologist J. B. S. Haldane, in *Possible Worlds and Other Essays*. He wrote:

> Gravity, a mere nuisance to Christian, was a terror to Pope, Pagan, and Despair. To the mouse and any smaller animal it presents practically no dangers. You can drop a mouse down a thousand-yard mine shaft; and, on arriving at the bottom, it gets a slight shock and walks away.
>
> A rat is killed, a man is broken, a horse splashes. For the resistance presented to movement by the air is proportional to the surface of the moving object. Divide an animal's length, breadth, and height each by ten; its weight is reduced to a thousandth, but its surface only to a hundredth. So the resistance to falling in the case of the

small animal is relatively ten times greater than the driving force. An insect, therefore, is not afraid of gravity; it can fall without danger, and can cling to the ceiling with remarkably little trouble.

JOHN FORRESTER

Don't Bee Home Late

The other day I noticed a large bee enter the door of my train carriage. Later, I saw it depart at a station 10 miles farther down the line. Is it likely the bee could find its way home, without using the train? If not, could it integrate into a new hive or colony, or would it face resistance and attack?

CHRIS BALL

Yes, there is a good chance the bee will find its way back home. Its large size suggests it was a queen or worker bumblebee—one of the *Bombus* species. Bees employ special orientation flights to memorize near and distant landmarks relative to the nest. As well as these cues, they also use the sun's position as a landmark, making use of a built-in clock to compensate for the sun's movement across the sky.

My colleague Mark O'Neill released tagged worker *Bombus terrestris* and *Bombus pratorum* bees at staged points from their nests and all returned safely, the farthest from 4 miles away. The bees were taken to their starting points by car, in the expectation that their usual navigational aids would be compromised, much as one might expect from a train journey. However, it is likely that their internal clock, compensating for the movement of the sun, enabled them to use solar positioning to fly back to a distance from the nest at which their visual cues would come back into play.

The ability to find a nest from a long distance away is vital to bees because nest sites and food may not be found in the same habitat. The very large females of the solitary, nonsocial bee genera *Anthophora* and *Proxylocopa*, whose nesting biology I studied at one site in the Negev desert, always flew straight out from the nest entrance and over the horizon formed by the shoulder of a hill about a third of a mile away. From the top of this hill, I was unable to see any suitable flowering vegetation in the next valley. I calculated that the minimum foraging distance from the nest site was 2.5 miles, requiring a minimum round trip of 5 miles.

It is also known that workers of the much smaller honeybee *Apis mellifera* can forage up to 8 miles from the hive, even in wooded country, and females of some large neotropical bees are thought to have foraging ranges of up to 20 miles

If the train-traveling bee was a queen *Bombus*, it might have been able to insinuate itself into an established colony of the same species by lying low in the nest to give it time to absorb the colony odor and avoid aggressive responses from resident bees. An egg-bound queen with aggressive tendencies might kill the resident queen and assume control of the colony. But an exhausted and disoriented bee entering a strange colony may well be killed by the workers. All of these behaviors have been reported for bumblebees.

A female *Anthophora* unable to find its nest might excavate a new nest if suitable sites were close by. However, as far as I know, no one has tried to see if such displaced solitary species will set up nest at a new site.

Chris O'Toole
Bee Systematics and Biology Unit, Oxford University
Museum of Natural History, Oxford, UK

Vicious Fruit

Why did pineapples evolve a fearsome array of spiny leaves that make the large, sweet, and juicy fruit almost impregnable? Surely the usual purpose of a sweet fruit is to encourage seed-dispersing animals to eat it. So what disperses the pineapple in its native range?

COLIN WILSON

The short answer is that pineapples aren't eaten in the state that we normally see them but are eaten after they have ripened much further and fallen to the forest floor.

The pineapple, *Ananas comosus*, was originally found in south Brazil and Paraguay and indigenous peoples spread it throughout South and Central America and to the West Indies.

The plant is a herbaceous perennial and grows up to 5 feet high and 3 feet wide. It has a rosette of long pointed leaves around a terminal bud. This bud produces the flowering stem which turns out an inflorescence of reddish purple flowers, each attached to the rest of the plant by a leaf-like structure called a pointed bract. In the wild these flowers may be pollinated by hummingbirds and will produce small, hard seeds in the fruit.

As everyone who eats pineapples knows, commercially grown fruit have no seeds. That is because pineapples, like bananas, will still develop fruit even if they are not pollinated and fertilized. Like many other plants, pineapples are unable to self-pollinate.

The pineapple fruit is created by the fusion of between 100 and 200 individual fruitlets that are embedded in a fleshy edible stem. The ovary of each flower becomes a berry, and all the berries coalesce into one solid structure. This is referred to as a multiple fruit or sorosis. The tough, waxy impregnable skin still contains the pointed bracts and the remains of the flower.

Although the pineapple plant can grow from seed, it also spreads very efficiently by a variety of vegetative means: from slips that arise from the stalk below the fruit, suckers that originate at the leaves, crowns that grow from the top of the fruits, and ratoons that come out from the underground portions of the stems.

The pineapple we buy today in the supermarket is very different from its natural relatives in South America. The wild pineapple is much smaller. By the time it has dropped off its stem, hit the ground from quite a height, and lain on the forest floor for a few days in the hot sun, it is very ripe and very soft. So when eaten, it is likely to be mushy and to split open easily, revealing the sweet and juicy fruit inside.

Humans tend to eat commercial pineapples and bananas before they are truly ripe. However, a soft, mushy pineapple lying on the ground is likely to be attractive to many animals including monkeys and small mammals, which will help spread the seeds.

Thanks to Philip Griffiths, Royal Botanical Gardens, Kew, London, for his help in compiling this answer.—Ed.

Flying V

I read a while ago that there are several competing theories as to why geese fly in a "V" shape. Does anyone know the definitive answer?

BRUCE SHULER

When the lead bird completes a flap of its wings two vortices are shed, one from each wing tip. These vortices consist of a rolling tube of air, the upper portion of which is moving forward and the lower part rearward. Should a following bird complete a downward stroke into the top of a vortex, the momentum change of the air caught up in the stroke is much greater than had the vortex not been present. Consequently the lift for a given stroke size is greater, and the following bird needs to do less work. To make use of this phenomenon the following two birds must be behind the wing tips of the lead bird in a V formation, and the birds behind them should be similarly placed.

This leads to an obvious question: why don't birds take up the position on the inside wings to form a tree formation? The answer is that they would be subject to vortices on both wings that were not synchronized, making flying difficult.

DAVID MANN

There are vortices in the air above, behind, and just outside the wingspan of any aircraft. The flow in the vortex region effectively sucks a following aircraft forward, and if the follower is properly positioned, it can gain additional lift from the upflowing region of the vortex.

Birds, too, are sensitive to any small variations in airspeed and direction that can help them fly. Birds flying in a group will drop back to take advantage of the wing vortices of their companions. The lead bird changes as each becomes tired, or when the formation alters course.

Competition glider pilots use the same trick. If you fly behind and to the side of an identical glider ahead of you, you will gradually catch up with it. If you are higher, you can overtake by converting your extra potential energy into speed at the last moment.

ALAN CALVERD

Flying in a V formation is more efficient for the following birds as the leader creates wing-tip vortices. NASA has been testing this idea with military aircraft. You can find it at www.nasa.gov/centers/dryden/home/index.html.

DOUGLAS YATES

My conclusion is that geese adopt the V formation as a result of several factors, none of which has anything to do with aerodynamics or wing-tip vortices.

The vortex theory seems to be based on what is known about vortices generated by fixed-wing aircraft. Flapping flight generates more complex vortices that are less well understood. One thing is certain: if a bird were to gain some advantage from the vortices of the one in front it would have to flap at the same rate and with a particular phase relationship to the flapping of the other bird in order to obtain the same effect on each wing beat. In fact, each bird settles into its own most comfortable flapping rate without reference to the rest of the flock.

Geese, whether migrating or just flying between their roosting and feeding areas, follow the lead bird at the apex of the V. This is in contrast to the behavior of starling flocks, in which each bird reacts to the movement of the birds closest to it in the flock. The leading bird determines the course and altitude of the flock, but other birds will give frequent inputs in the form of honking or stretching the line. Sometimes a bird will leave the main flock and lead a breakaway formation on a divergent course. Usually a compromise is reached and, after a lot of honking, the flock reforms, often with a different leader. For this behavior to work each bird has to follow the one in front, so random positioning or flying abreast in a line ("line abreast") is no use. They must therefore take up "line astern" or echelon formation.

A goose's eyes, like those of most prey species, are set in the sides of its head, giving good all-around vision but leaving a small blind spot directly ahead and behind. If a goose were to follow directly behind the one in front it would have to turn its head slightly to see the other one clearly and would have to resort to asymmetrical flapping to maintain a straight course, reducing its aerodynamic efficiency and wasting energy. It would also have to fly a little below the one in front in order to stay clear of the wake, not a good place to be as geese defecate in flight. This leaves some kind of echelon formation as the only practical solution.

CHARLIE BATEMAN

The explanations given above by the first three authors and by Charlie Bateman certainly seem contradictory. Indeed, there has been a long-running argument between the supporters of the "aerodynamic" and the "behavioral" explanations for V-shaped flight. But there is good reason to believe that both are true.

There are many observations like those given by Bateman of birds coordinating their flight path and following one another. A V-shaped formation does allow all birds to keep a close eye on one another and makes it harder for predators to single out any one bird for attack.

But there is also experimental evidence that birds flying in this formation use less energy. Experiments reported in Science *in 1970 by P. Lissaman and C. Shollenberg (vol. 168, p. 1003) showed that geese flying in a V can travel 70 percent farther than solo birds. More recently a group of French researchers led by Henri Weimerskirch directly measured the wing beats and heart rates of great white pelicans flying in and out of formation* (Nature, *vol. 413, p. 697). Their ingenious experiments were carried out in the Djoudj National Park in Senegal and used birds that had been trained to follow an ultralight airplane or motorboat. The birds were filmed as they flew and heart-rate monitors were strapped to their backs. The researchers found that formation flight does indeed provide a significant aerodynamic advantage, partly by lengthening the time for which they can glide (see Alan Calverd's reply above).*

Pelicans do not always fly at the optimum distance from one another to maximize their energy

saving, and other birds fly in flocks where there is no aerodynamic advantage, or possibly a disadvantage compared with solo flight. Even geese often fly in straggling formations that do not achieve the maximum possible energy saving.

Putting these observations together suggests that there is more than one advantage to flocking, and that aerodynamic and social benefits may have evolved together.—Ed.

Dem Bones

Fluffy, my pet guinea pig, died recently. We buried him in a cardboard shoe box about 30 inches down. I am driving my mom mad by asking if Fluffy is just bones yet. Is he? And if he isn't, when will he become just bones? And if he doesn't, what will happen to him?

—DIMITRI

Fluffy was buried four weeks ago in the San Francisco Bay area, which has a temperate climate. I am constantly being asked by Dimitri, "Is Fluffy bones now?" I would appreciate some detailed information on decomposition and the time involved in order to help me answer the question. The topic may sound a bit gruesome for an 8-year-old but my son thrives on science.

—MOM

Dimitri Maxwell (age 8) and Kathleen Wentworth

Dear Dimitri's Mom,

Thank you for sending Dimitri's very interesting question to *New Scientist*. My husband and I are field biologists who teach at a college, run a consultancy, and also have a farm not far from the San Francisco Bay area. Dimitri's inquiry didn't seem in the least strange or macabre to me. Teaching has shown me that elementary school boys have a great interest in all aspects of the decomposition side of the food chain.

Dear Dimitri,

It's a very hard one to answer precisely. My husband and I are biologists and longtime farmers in your area. Farmers, especially those who raise livestock, bury a lot of small animals over the years. We have also prepared many animal carcasses or parts of them as study specimens. And yet we still don't always know what to expect when an animal starts to decompose.

There are a great many variables, different factors which change the time it takes for all the flesh and skin and fur to leave the bones. My guess is that Fluffy won't be "just bones" for at least six more months, and because he was buried pretty deeply and in a box, I'd expect the whole process to take a year or more.

This is based on what happens to gophers and rats, animals about the size of Fluffy, when we bury them here, and we've buried many over the years. The body will be worked on by a huge number of organisms whose job it is to return plant and animal tissue to the soil. They are of all kinds and all sizes and, as a group, are known as decomposers. I outline some of their details and characteristics below.

Bacteria and fungi that live in the soil and within the body itself will work on the soft parts of the animal, eventually returning the elements to the earth to nourish plants and soil organisms. There are three obvious variables in this part of the cycle.

The first is that some soils have a lot more bacteria than others. Second, most of these microorganisms need oxygen. The deeper you go into the ground, the less oxygen there is in the soil, and I see that Fluffy is buried pretty deep.

Finally, most of the decomposer microorganisms like it warm and moist. Your soil is probably only medium-warm at 30 inches down—probably something around 56°F or a little cooler would be my guess. If Fluffy's body is in your garden, it's probably nicely moist, but if he's in unwatered soil the decomposers' work will be slowed down quite a lot, because I know our region hasn't received enough rain to penetrate more than an inch or two this summer.

If I bury a gopher in summer in this area, not as deep as Fluffy is buried and not in a box, it won't have become just bones if I accidentally dig it up when working in my garden the following February, even though all the factors I list above are at a high level here. That's why I estimate that it would be at least another six more months before decomposition occurs fully.

Some larger soil invertebrates may be able to help the box disintegrate and also go to work on the body. Some eat meat, and some, such as dermestid beetles, specialize in skin and hair. Vertebrate museums which prepare study specimens of skeletons actually keep colonies of these beetles, to finish cleaning the bones. They are also found fairly frequently on livestock farms like ours, and in woodland where they help to clean up casualties.

And there are other invertebrate meat-eaters that live in the soil, such as ants and worms and beetles. There are more in the upper layers than at the depth where Fluffy is buried, but some may get down that far.

There are a few other points. Eventually, even the bones will disintegrate and feed the plants and the soil, although Fluffy's teeth will be around the longest—tooth enamel is a very durable material.

Also, the soil organisms often move bones around as they work on them. If you dug him up, I wouldn't expect to see a perfect skeleton of Fluffy, all in one place, though you might be lucky because he has a box around him.

There is actually a center on the East Coast of the

United States that is devoted solely to studying how animal and human carcasses decompose and then disintegrate. The forensic anthropologist Douglas Ubelaker describes it in his book *Bones*. It seems that even scientists who specialize in old bones still don't know as much about the process as they would like to.

MOLLY

The amount of time that it takes for an animal like Fluffy, or even a human, to decompose to the point of becoming skeletonized is highly variable and depends on many factors.

Those factors include, but are not limited to, the season of the year, the ambient temperature, the amount of rainfall, the depth of burial, whether insects can reach the body, the pH of the soil, and whether the body was embalmed or not.

The size of the body also plays a part in determining how quickly decomposition occurs. Additionally, if a body is enclosed in a coffin, or wrapped in plastic bags or carpet, for example, it will take longer to decompose because it is protected. In Fluffy's case, the depth of burial at approximately 30 inches, and the fact that he was buried in a cardboard shoe box, would slow the process of decomposition. It is probable that Fluffy has not yet become just bones.

LESLIE EISENBERG
Board-Certified Forensic Anthropologist
State Historical Society of Wisconsin,
Madison

In the Dock

Why are dock leaves so effective at relieving stings from nettles? Are they effective on any other plant or insect stings?

TIM CROW

Being stung by a nettle is painful because the sting contains an acid. Rubbing the sting with a dock leaf can relieve the pain because dock leaves contain an alkali that will neutralize the acid and therefore reduce the sting. Bees and ants also have acidic stings, so dock leaves should help, but other alkalis, such as soap or bicarbonate of soda, are usually better.

However, a dock leaf is useless against wasp stings, which contain an alkali. This is unfortunate because wasps are nasty little critters whose sole aim in life is to ruin picnics and barbecues. If you want to neutralize a wasp sting you should use an acid such as vinegar. The only problem is that you'll smell of pickles for the rest of the day.

PETER ROBINSON

A Sting in the Mouth

In a recent conversation about food chains, a colleague wondered if anything ate wasps. Someone suggested "very stupid birds." Does anyone know any more about this?

TOM EASTWOOD

The lowly wasp certainly has its place in the food chain. Indeed, the question should possibly be, "What doesn't feed, in one way or another, on this lowly and potentially dangerous insect?"

Here are a few creatures that do, the first list being invertebrates: several species of dragonflies (Odonata); robber and hoverflies (Diptera); wasps (Hymenoptera), usually the larger species feeding on smaller species, such as social paper wasps (*Vespula maculata*) eating *V. utahensis*; beetles (Coleoptera); and moths (Lepidoptera).

The following are vertebrates that feed on wasps: numerous species of birds, skunks, bears, badgers, bats, weasels, wolverines, rats, mice, and last, but certainly not least, humans and probably some of our closest ancestors.

I have eaten the larvae of several wasp species fried in butter, and found them quite tasty.

ORVIS TILBY

The definitive source on European birds, *Birds of the Western Palearctic*, lists a remarkable 133 species that at least occasionally consume wasps. The list includes some very unexpected species, such as willow warblers, pied flycatchers, and Alpine swifts, but two groups of birds are well-known for being avid vespivores. Bee-eaters (Meropidae) routinely devour wasps, destinging them by wiping the insect vigorously against a twig or wire. And honey buzzards raid hives for food. They are especially partial to bee larvae, but in the United Kingdom, wasps, again mostly larvae, also form a major part of their diet.

SIMON WOOLLEY

I have a photograph taken in my garden, showing a mason wasp having its internal juices removed via the proboscis of a large insect.

TIM HART

In July 1972 I was snorkeling off the Californian island of Catalina. I returned to the east cliff of the island as sunlight was leaving the shore. In a crevice at the base of the cliff I saw a crab holding a wasp, which was still moving.

I took a photograph which shows the right pincer holding part of the wasp while the left pincer carries the wasp's abdomen to the crab's mouth.

The crab did not show any sign that it was startled by the taste of its meal.

GARRY TEE

Badgers will dig out a wasps' nest and eat the larvae and their food base. During the summer of 2003 I saw an underground nest being demolished by badgers.

TONY JEAN

I was once idly observing a wasp crawling around the edge of a water lily leaf in my pond when it paused to drink. There was a sudden flurry of activity when a frog leaped from its hiding place and swallowed the wasp.

The frog did not appear to suffer any ill effects, so I captured another wasp, tossed the hapless creature into the pond, and waited. The frog was slow on the uptake, but there was another disturbance in the water and this time a goldfish snapped up the wasp. The fish, too, seemed undisturbed.

My curiosity now thoroughly aroused, I wondered whether the fish could be induced to consume further wasps. For the next hour or so I continued to hunt down luckless wasps and throw them into the pond. Some got away, some were eaten by the fish, and a few were swallowed by the frogs.

JOHN CROFT

Returning home late one night I heard the persistent buzzing of a wasp in the kitchen window. It appeared to be struggling around at the bottom of the window, unable to fly properly. A tiny red spider was attached to the underside of its abdomen. The spider must have been some 20 times smaller than the wasp and was positioned where the wasp was unable to mount a counterattack.

The next morning revealed an empty, transparent wasp exoskeleton.

JOHN WALTER HAWORTH

3 Domestic Science

Bluto Strikes Back

My Italian recipe book says that I should cut cooked spinach with a stainless steel knife to avoid discoloration. Which would become discolored if I don't—the knife or the spinach? And what would be the chemistry at work?

HANS HAMICH

The reason you should you always use a stainless steel knife to cut your spinach is intriguing and is a major stumbling block to the fortification of food with iron. Remember that a lack of iron is the world's most prevalent nutritional deficiency.

Both iron blade and spinach will become discolored because of the reaction between the polyphenols in the spinach and the iron blade. If you want to see a dramatic illustration of this, make yourself a cup of tea and add a few crystals of a soluble iron salt such as ferrous sulfate (don't drink it).

The black discoloration you see is caused by the reaction between polyphenols, called tannins, in the tea, and the iron. The resulting black compound is highly insoluble. The implications for iron absorption in the body are huge because iron in this form is virtually unavailable for absorption. So whatever the source of strength in Popeye's spinach, it is not the iron.

Polyphenols are found in many vegetables and, together with phytates, are the reason why many people who subsist on cereal and vegetable diets have an iron deficiency. Fortifying such diets with iron salts creates two problems. First, the iron is not absorbed; and second, the colored iron-polyphenols make the food look unattractive.

PATRICK MACPHAIL
Department of Medicine
University of the Witwatersrand
Johannesburg, South Africa

Beer Orders

I occasionally add lemonade or ginger beer to a glass of beer to make a shandy. If the beer is poured first and the lemonade or ginger beer is added afterward, the contents will fizz and even overflow. Pouring the soft drink first and then topping it up with beer avoids the problem. Why is this?

BRYAN HARRIS

If you pour the liquids into separate glasses you will notice that only the beer forms a head. That is because it contains surfactants, proteins, and other long-chain molecules that help liquid films to form and stabilize bubbles. Lemonade bubbles, on the other hand, burst much too quickly to form a head.

Lemonade that is poured into beer sinks turbulently, stimulating vigorous bubble formation. The beer on top is barely diluted at first, so bubbles surfacing through it quickly form a beery froth. Beer poured into lemonade also rushes to the bottom, but in this case the bubbles surface in lemonade, and therefore quickly pop as usual. By the time enough beer to support a froth has reached the top, most of the fizzy fuss is over and so no head forms.

A possible extra cause of shandy foam's weakness is that

liquid films are terribly sensitive to unequal strengths in their surfaces. Mix practically any two different kinds of foam, and the bubbles that are present start breaking much faster. As an experiment, pour a glass of beer with a good head, then add droplets of lemonade, dishwashing liquid, gin, grains of salt, squirts of citrus zest, and so on, and see which of them causes the most drastic erosion of the froth.

JON RICHFIELD

Spectral Images

When condensation forms on a clean bathroom mirror, you can draw pictures in it. When the condensation evaporates, the pictures disappear. But when it forms again, they reappear. Why?

GLYN WILLIAMS

When water vapor condenses on a dry mirror, it does so as separate droplets, a process known as dropwise condensation. The numerous drops effectively screen the mirror so that it appears opaque.

When you draw on the surface with your finger, the droplets coalesce into a thin film of transparent water, so the mirror becomes reflective again in these areas. When the mirror warms up, or the air humidity falls, the droplets evaporate. The image disappears because the surrounding droplets no longer contrast with it.

The film of water evaporates more slowly than the droplets because of its lower surface area. If it does not have time to completely evaporate, any condensation occurring soon afterward will be dropwise where there were droplets before, and in a film where some of the film remains. This latter process is known as filmwise condensation. The image then reappears on the glass.

If the mirror dries completely, the pattern should not normally reappear when further condensation occurs, though it might if the surface has been contaminated where you drew the image. Drawing a finger across the mirror in making the image may leave traces of sweat on the surface which would, owing to its salt content, help to promote filmwise condensation.

Dropwise condensation is known to chemical engineers to be more efficient at transferring heat than filmwise condensation, but in practice it is much more difficult to promote, because as the droplets enlarge, they touch each other and coalesce, so the process tends to become filmwise. On the other hand, dropwise condensation is easy to prevent.

Wiping the mirror with a cloth or a tissue wetted with a small amount of detergent such as shampoo leaves an invisible film on the surface. This reduces the surface tension of the condensing droplets, causing them to flatten out and readily coalesce into a film. This is the basis of the antimisting fluids which are used for treating spectacle lenses and car windshields.

TONY FINN

When you draw an image in the condensation mist, you leave traces of finger grease (or, if you have just washed, grease plus shampoo or soap). The film is transparent, so you don't see it when the condensation clears. The next time water vapor condenses on the cold mirror, there is a difference in droplet size between condensation on clean glass and on contaminated glass.

In some cases, it is the contaminated glass that encourages droplet formation, and then you see the image as positive rather than negative. But usually water-loving surfactants such as soap reduce the formation of droplets and generate a smoother, clear film of water, contrasting with the gray mist on the surrounding glass.

HUGH WOLFSON

Whisking Disaster

For years, whenever my family has come to visit, I have made meringues, which involve whisking egg whites until they are thick. I have always used free-range eggs, but recently I bought organic free-range ones, and no matter how much I whisked these whites they would not thicken. Why should organic eggs behave in this way? Is something missing from the birds' organic diet that prevents the whites of their eggs from thickening?

VERA GAYLOR

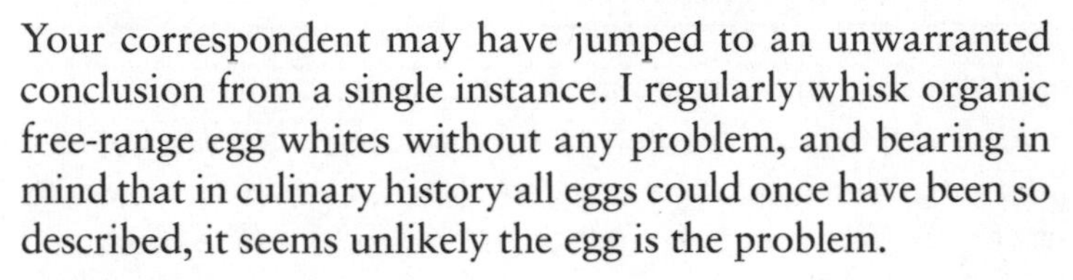

Your correspondent may have jumped to an unwarranted conclusion from a single instance. I regularly whisk organic free-range egg whites without any problem, and bearing in mind that in culinary history all eggs could once have been so described, it seems unlikely the egg is the problem.

J. OLDAKER

A good, stiff meringue froth demands a complex interconnection of suitably distorted protein molecules. Anything that interferes with the interlinking of the molecules leaves the whites an unappetizing slush. Oil is the usual culprit. Use clean, dry utensils, free of detergent. Before the whites have formed a stiff froth, the merest drop of cooking oil, cream, or oily yolk that gets into the whites can ruin the meringue.

JON RICHFIELD

The eggs from our pet chickens, while not certified organic, are as close as we can get, and make wonderful meringues. The problem, I suspect, is freshness. Whites from eggs less than 5 or 6 days old will not whip up. Supermarket eggs can be up to 2 weeks old, but the organic ones were probably newly laid.

This raises the question of what changes eggs undergo as they age to allow them to make successful meringues.

PHIL BAKER

The eggs may have been too young. I assume that the protein molecules develop cross-linkages as the egg ages, enabling the albumen to contain the air bubbles when whipped.

LORNA ENGLISH

Concerned Consumer

I always use blue toilet paper because it matches my bathroom decor. However, a friend told me that I should use only white, because colored paper is more damaging to the environment. My local supermarket sells a huge variety of colors with any number of patterned varieties too. Is it true that some varieties are more environmentally damaging? And if so, why? Are paper towels even worse than toilet paper?

JOHN SHAW

If your friend means that the dyes are ecologically harmful, forget it. Chemically active groups on the dye molecules cling to the cellulose; that is why the colors don't run and leave you fundamentally decorative after you apply them. The dyes are like a mousetrap that has caught a mouse: the trap, in demonstrating its bite, has become harmless. Much as the trap is hard to reset, the dyes are hard to release from the paper.

Dyes are expensive, and toilet paper requires only traces, so even the most environmentally unaware manufacturer will prefer safe dyes that are simple to handle, and can be applied stingily, typically in parts per million. When the paper reaches the sewage works, the immobilized molecules soon succumb to bacteria, so they do not accumulate in the environment.

If you doubt this, buy a job lot of toilet paper; fold wads of, say, 10 squares, each of a single color; bury them separately in moist garden soil; and in a month or two exhume them and observe the result. In good soil you will do well even to detect your test pieces after the earthworms have done their work.

Much the same applies to paper towels, except that their strength while wet may mean they break down more slowly. Their persistence probably does more to provide bacteria with a durable home than harms the environment in any way.

Anyway, what about the bleaches necessary for producing white toilet paper? If you really want to be politically correct, go for garbage gray.

JON RICHFIELD

Wood is brown. Unbleached paper is brown. White paper improves the contrast between text and background to aid reading, so most people prefer it. To make paper white, it is usually bleached with chlorine, which can form carcinogenic dioxins. The paper industry has substantially cut down the quantity of dioxin by-product it produces, and there are initiatives to eliminate it totally by using only hydrogen peroxide and ozone bleaches, which are somewhat more expensive.

Incidentally, what we consider brilliant white is actually slightly blue. Many papers therefore contain fluorescent whitening agents (FWAs), that reemit UV light as blue light, plus some blue dye. You may have seen clothes and paper containing FWAs glowing under "black lights."

BRADY HAUTH

Pickled Poser

I love pickles and chutneys. But I'd like to know what nutrition is conserved and what is lost when vegetables are prepared and preserved in this manner.

AIDAN HANCOCK

Pickles and chutneys were originally a means of preserving fruit and vegetables using a combination of heat processing to kill bacteria, fungi, and yeasts, with added sugar, acid (in the form of vinegar), and salt acting as preservatives.

In some cases the raw fruit or vegetable is fermented for weeks or months in a brine containing lactobacillus bacteria, which produce the natural preservative lactic acid. Mango and cabbage can be preserved in this way. This also develops the texture and flavor of the fruit or vegetable.

Vitamins, minerals, and nutrients are lost when the cooked or fermented plants are washed during processing. Unstable vitamins start to break down soon after harvesting; this breakdown process is accelerated at high cooking temperatures and at the acidic pH typical for these types of product.

Some pickles and chutneys are prepared in oil rather than a sweet sauce, and there is generally less vitamin breakdown in the oily type because cooking is less severe. If we take mango as an example, the raw fruit contains about 40 milligrams of vitamin C per 4 ounces of fruit, while oily mango chutney contains only about 1 milligram and in sweet chutney vitamin C is barely detectable.

MARK WAREING

Many pickles are lightly cooked, or even just blanched or fermented. But chutneys are nutritionally different, as they are cooked almost as aggressively as jam. Both are products of pre-refrigerator preservation technologies, but their preparation and storage cause nutrient loss in four main ways: leaching, heating, oxidation, and degradation. The main losses are of soluble or unstable nutrients such as some vitamins, antioxidants, and minerals.

Pickling fluid itself causes hardly any degradation. How much leaching occurs depends on how the pickles are cooked and stored in liquid. For instance, large chunks leach less than grated material, which has a larger surface area. Using pickle fluid in soups or stews is a tasty way to reduce this loss. Bulk nutrients such as starches and proteins are not much affected, and in fact processing may improve their digestibility.

In modern pickling mild preservatives prevent decay. Manufacturers also rely on opened containers, being kept cold to slow degradation and decay. Darkness also protects light-sensitive vitamins such as A and C. To prevent oxidation, jars of preserves should be closed tightly and used soon after being opened.

Pickles have an honorable nutritional history. As well as simply combating starvation, the likes of sauerkraut have prevented many a case of scurvy during northern winters.

ANTONY DAVID

Dunking Dumplings

Whenever I am preparing Italian potato dumplings, or gnocchi, I notice that they behave strangely. When I put the frozen gnocchi into lightly salted boiling water, they immediately sink to the bottom. But the main ingredient of frozen gnocchi is frozen water, whose density is about 62.4 pounds per cubic foot, and the density of boiling water is 59.8 pounds per cubic foot, so shouldn't the gnocchi stay afloat until the ice melts and then sink to the bottom? Instead, they rise to the

surface after 2 minutes and all float when cooked, when they should be heavier than water. What is going on?

Radko Istenic

When the frozen gnocchi are placed in hot water the combined density of all the ingredients is greater than the density of boiling water, and therefore the gnocchi sink. As the gnocchi warm up it's a bit like inflating a rubber dinghy at the bottom of a swimming pool. The air trapped in the dough expands and the combined density of all the ingredients becomes less than the density of boiling water, causing the gnocchi to rise to the top.

Martin Garrod

I've been making gnocchi for a long time and, as I had some frozen ones, I decided to do some rudimentary measurements in my kitchen.

First, my frozen gnocchi had a density of 0.63 ounce per cubic inch and they duly sank in plain boiling water. When they came to the surface and were well cooked, I scooped them up, drained them on a towel, and took the same measurements again. This stage was very fiddly so the results must be taken with a pinch of salt. There was a 14 percent increase in volume; there was an 8 percent increase in weight; and their density was reduced by 5.5 percent. Curiously, when I placed them in cold tap water to measure the volume, my cooked gnocchi sank.

Gnocchi sink because they are denser than water. However, the dough of well-made gnocchi has many small bubbles of air which stay there and expand when the gnocchi are placed in boiling water, so they come to the surface. My cooked gnocchi sank in cold water because the air trapped inside contracted slightly.

Maria Fremlin

Spice Attack

Why does turmeric stain everything an indelible yellow, including surfaces that appear impermeable to other substances? Other powdered spices such as cinnamon, paprika, and chili do not leave the same legacy. And what is the best way to remove turmeric stains?

HEFIN LOXTON

Turmeric, the powdered rhizome of *Curcuma longa*; and paprika, which is obtained from the fruits of sweet peppers, *Capsicum annuum*, are examples of spices used in cooking as much for their color as for flavor.

The yellow colour of turmeric is caused by curcumin, which makes up around 5 percent of the dry powder. The red pigments in paprika are a mixture of carotenoids, principally capsanthin and capsorubin, and in dried paprika they amount to a maximum of 0.5 percent of the weight.

The red carotenoids, which consist of long, chain-like molecules, are soluble in organic solvents such as petroleum spirit. Curcumin consists of smaller molecules with terminating phenyl groups. It is insoluble in water but dissolves in solvents like methanol. Therefore you might expect that both paprika and turmeric would stain paintwork and plastics, because they dissolve in organic solvents. You would also expect them to migrate to the oily part of food during cooking.

To compare their coloring properties, put a good pinch of turmeric into two small glass spice jars and do the same with paprika. Add a dessertspoon of methylated spirit to one set and the same amount of white spirit to the other (you can repeat the experiment with cinnamon and chili powder). Upon shaking the mixtures you will see a vivid yellow color appear instantly in the meths from the turmeric powder and the white spirit turn red from the paprika. When you put a drop of extract from each of the four jars onto a clean white

plate, you will see that the turmeric-meths extract has much the strongest color, followed by the white spirit and paprika. The same experiment can be done with acetone (nail varnish remover).

This demonstrates the principal reason why turmeric stains more than other spices—it simply has more extractable coloring material in it. Other reasons will reflect the different physical properties of curcumin and the red carotenoids, as demonstrated in our solubility experiment, and differences in the way the dyes react chemically with solid materials. Drops of your extract can be placed on various surfaces to test their staining ability, but do ask first.

Curcumin is stable when heated but is not stable when exposed to light. So to remove a turmeric stain, first clean with methylated spirit and then place the object in sunlight.

MICHAEL ELPHICK

Rubber Horror

Why do rubber bands spontaneously melt? Often I find an aging one on my desk that has turned into a sticky mess. After a few more months, the sticky mass solidifies and becomes brittle. Why?

STUART ARNOLD

Natural rubber is made of polyisoprene chains that slip past each other when the material is stretched. When raw, the substance is too sticky and soft to be of much use, so it is toughened with the addition of chemicals such as sulphur that create cross-links between the chains, making the rubber stiffer and less sticky. This process is called vulcanization.

With time, ultraviolet light and oxygen in the air

react with the rubber, creating reactive radicals that snip the polyisoprene chains into shorter segments. This returns the rubber to something like its original state—soft and sticky. Meanwhile, these radicals can also form new, short cross-links between chains. This hardens the rubber and eventually it turns brittle. Any vulcanization agents left in the rubber contribute to the process.

Whether a rubber band goes sticky or hard depends on the relative rates of these processes, and these rates in turn depend on the rubber's quality such as what additives, fillers, and dyes it contains—and how it is stored. Heat and light speed up the reactions (for example, an 18°F rise in temperature will roughly double reaction rates), and the presence of strong oxidizers such as ozone creates even more radicals. The eventual fate of your rubber band depends on the temperature in the room, and on whether you have a desk by a window or near a machine such as a photocopier that creates ozone.

How much light and heat is required for these changes? The polymer chemistry of rubber is fairly messy, and so this is difficult to answer precisely. Obviously, the chemical reactions run slowly if the rubber is in a refrigerator, more quickly if left on a sunny desktop. A rule of thumb is that reaction rates roughly double for an 18°F rise in temperature, but this is complicated when you take oxygen and light into consideration. The quality of the rubber is also important, such as whether it contains additives, fillers, or dyes that absorb light energy or help transfer radicals. The final factors that influence the change are ozone concentration, UV light intensity, and whether the band is stretched or not—stretching brings chains closer together, allowing radicals to jump from one chain to another more easily, and to create new bonds between chains.—Ed.

Citric Secret

Why does lemon juice stop cut apples and pears from browning?

BRIAN DOBSON

To answer this question we first need to understand why some plant tissues go brown when cut. Plant cells have various compartments, including vacuoles and plastids, which are separated from each other by membranes. The vacuoles contain phenolic compounds, which are sometimes colored but usually colorless, while other compartments of the cell house enzymes called phenol oxidases.

In a healthy plant cell, membranes separate the phenolics and the oxidases. However, when the cell is damaged—by cutting into an apple, for example—phenolics can leak from the vacuoles through the punctured membrane and come into contact with the oxidases. In the presence of oxygen from the surrounding air these enzymes oxidize the phenolics to give products that may help protect the plant, favoring wound healing, but also turning the plant material brown.

The browning reaction can be blocked by one of two agents, both of which are present in lemon juice. The first is vitamin C, a biological antioxidant that is oxidized to colorless products instead of the apple's phenolics. The second agents are organic acids, especially citric acid, which make the pH lower than the oxidases' optimum level and thus slow the browning.

Lemon juice has more than 50 times the vitamin C content of apples and pears. And lemon juice, with a pH of less than 2, is much more acidic than apple juice, as a quick taste will tell you. So lemon juice will immediately prevent browning.

You could also prevent cut apples from browning, even without lemon juice, by putting them in an atmosphere of nitrogen or carbon dioxide, thus excluding the oxygen required by the oxidases.

An excellent vegetable for observing browning is celeriac. It is possible to cut a large, relatively uniform slice of this root tissue and then lay several small filter paper disks on the cut surface, each soaked in a different solution such as lemon juice, apple juice, vitamin C, other antioxidants, citric acid, other acids, and suchlike. A disk soaked with an agent that blocks the action of oxidases will leave a white circle on an otherwise brown surface.

STEPHEN C. FRY
Institute of Cell and Molecular Biology
University of Edinburgh, UK

Polyphenol oxidase (PPO) was discovered in mushrooms in 1856 by Christian Schoenbein. It is widespread in nature and found in humans, most animals, and many plants. In plants its function is to protect against insects and microorganisms when the skin of the fruit is damaged. The dark brown surface formed by the skin is not attractive to insects or other animals, and the compounds formed during the browning process have an antibacterial effect.

In some foodstuffs made from plants this browning effect is desirable. For example, in tea, coffee, or chocolate it produces their characteristic flavor. However, in other plants or fruits such as avocado, apples, and pears, browning is an economic problem for farmers, because brown fruit is not acceptable to consumers and it doesn't taste good.

ANGELES HERNÁNDEZ Y HERNÁNDEZ
Laboratory of Crystallography
Andalusian Institute of Earth Sciences, Granada, Spain

The Black Stuff?

When I buy a pint of Guinness there is no doubt the liquid is black. Yet the bubbles that settle on top, which are made of the same stuff, are white. The same is true of many types of beer. Why?

STEWART BROWN

In the interest of science I poured myself a Guinness and waited until the rising bubbles had formed a creamy head. I put a little of this in a dish and examined it through a low-powered microscope. Unlike bath foam, which has many semicoalesced bubbles, Guinness foam is made mainly of uniformly sized, spherical bubbles of about 0.1 to 0.2 millimetres in diameter, suspended in the good fluid itself.

Near the edge of the drop of foam it was possible to find isolated examples of bubbles, and by viewing objects held behind these it was clear that they were acting as tiny divergent lenses. Just as a clear spherical marble, which has a higher refractive index than the surrounding air, can act as a strong magnifying glass, so spherical bubbles in beer diverge light because the air they contain has a lower refractive index than the surrounding fluid.

As a result, light entering the surface of the foam is rapidly scattered in different directions by multiple encounters with the bubbles. Reflections from the bubbles' surfaces also contribute to this scattering. Some of the light finds its way back to the surface and because all wavelengths are affected in the same way we see the foam as white. Light scattering from foam is akin to the scattering from water droplets that causes clouds to be white. This is called Mie scattering.

I sat back and drained the glass. On closer inspection, the head of Guinness is actually creamy colored, and a drop or two that remained in the bottom of the glass had a light brown color. Although bulk Guinness appears black, it is not opaque. In the foam there is not so much liquid—most of the space is taken up by air. But because light is scattered from bubble to bubble the intervening brew does absorb some of it, providing a touch of color.

Needless to say, to ensure reproducibility the experiment was repeated several times.

Martin Whittle

Light bite

Aero, a famous brand of chocolate bar, contains bubbles in a chocolate matrix. The bubbles are evenly sized and distributed throughout the whole bar. How do the manufacturers produce this effect? Why don't the bubbles rise to the surface as the chocolate solidifies?

Natasha Thomas

The way in which the unique Aero bubbles are added is a top-secret process closely guarded by Nestlé Rowntree. We can tell you, however, that there are approximately 2,200 bubbles in one Aero chunky bar!

Marie Fagan
Press and PR officer, Nestlé UK

Secret details may be absent but the broad answer is in Rowntree's British patent GB 459583 from 1935.

The chocolate is heated until it is in a fluid or semifluid state, then it is aerated, for example using a whisk, to produce many tiny air bubbles distributed throughout the chocolate. This is poured into molds and the air pressure greatly reduced as the chocolate is cooled. The reduced air pressure causes the tiny bubbles to grow and gives the finished chocolate its frozen bubbles appearance. The solid chocolate coating on the surrounding surfaces of the bar is

placed into the mold before the aerated fluid chocolate is poured in.

The patent gives no clues on how the bubbles are prevented from rising to the surface during manufacture, but this may be due to the high viscosity of the semifluid chocolate and the rapid rate of cooling.

Patents provide a great source of technical information. It has been suggested that 80 percent of technical disclosures appear in patents and nowhere else. You can view and print GB 459583 using the service on the Patent Office website, www.patent.gov.uk. The service provides an interface to British and European patent offices for you to search their databases.

Melvyn Rees
Marketing and Information Division
The Patent Office, London, UK

It's not the chocolate answer that your reader was looking for, but I was once told that a soap manufacturer used the same process to make floating soap. The experiments were a technical success, inasmuch as the soap floated, but the product was not commercially viable because it dissolved too quickly.

Mike Dignen

David Bailey of Brookes Batchellor patent attorneys in Tunbridge Wells, Kent, UK, also picked up on another patent, GB 459582, and so did Armen Khachikian of the British Library's patents information section. This was filed by Rowntree on the same day as the one mentioned above and contains the Aero concept. The chocolate makers clearly knew what they were about. Khachikian points out that eight days before the patent was lodged, the Aero name was trademarked. Although British patents expire 20 years after they are filed, the trademark on the name Aero is still in force.

Thanks also to Tom Jackson of Wigton, Cumbria, UK, for ferreting out US patent 4272558 and British patent GB 480951 from 1938 on "Improvements in Confections for Eating or for Making Into Beverages" filed by Sydney Phillips and Arthur Whittaker. This patent contains information on making bubbles in molten chocolate with pressurized gas and then discharging the gas through a nozzle. It states: "The releasing of the chocolate into a region at atmospheric pressure causes the gas to escape from or expand within the chocolate, giving the chocolate a porous, cellular, honeycomb-like open structure."—Ed.

Cream On

One of the ways of drinking the liqueur Tia Maria is to sip it through a thin layer of cream. If the cream is poured onto the surface of the drink, to a depth of about 0.1 inches, and left to stand for about two minutes, the surface begins to break up into a number of toroidal cells. These cells develop a rapid circulation pattern which continues even if some of the Tia Maria is sipped through the cream. How and why do these cells develop and what is the energy source?

GEOFFREY SHERLOCK

We are glad to see that this Last Word question from 1995 inspired a research project by Julyan Cartwright at the Laboratory for Crystallographic Studies in Granada, Spain; Oreste Piro at the Mediterranean Institute of Advanced Studies in Majorca, Spain; and Ana Villacampa at the Lawrence Livermore National Laboratory in California. Their paper "Pattern Formation in Solutal Convection: Vermiculated Rolls

and Isolated Cells" was published in Physica A, *(vol. 314, p. 291). Over the years since 1995, a number of theories were sent to the* New Scientist *offices and the consensus was a reaction between alcohol and fat in the cream was responsible. Now we know the real answer, and the first author of the* Physica A *paper has sent us the following account—Ed.*

We tried this and were hooked. It is beautiful to watch these patterns form, and how different patterns form in layers of cream of different thickness.

This is all caused by convection. Convection is the bulk movement of fluid, often associated with temperature differences, called thermal convection. In Tia Maria and cream the convection is driven by a difference in concentration, and is called solutal convection.

The important component is the alcohol in the Tia Maria. After the cream is poured on top of the liqueur, the alcohol begins to diffuse through the cream layer. When it reaches the surface it alters the surface tension: the more alcohol at the surface, the lower the surface tension. Regions of higher surface tension then pull liquid towards them from the regions of low surface tension. As the surface liquid is pulled away, the liquid beneath these regions of low surface tension takes its place.

But this liquid contains yet more alcohol, because it has come from the part of the cream nearer to the Tia Maria below. It has an even lower surface tension and in turn gets dragged away. This positive feedback mechanism creates convection, which continues as long as there is a concentration difference to sustain it.

This type of convection, driven by surface tension, is called B'enard-Marangoni convection and it is particularly relevant to thin layers of fluid. It is important in situations like drying paint, and the same capillary or surface-tension forces also cause other types of patterns in alcoholic drinks, like the tears in glasses of wine.

The other important mechanism that can cause convection is buoyancy. But buoyancy-driven, or Rayleigh-B'enard convection, cannot be causing the patterns in Tia Maria because cream is lighter than Tia Maria, so Tia Maria with cream on top is buoyantly stable.

The patterns that form when a fluid starts to convect by either of these mechanisms have been well studied in situations such as rolls of clouds in the sky, or hexagons formed in a frying pan when a layer of oil is heated. Tia Maria is an oddity because the patterns are not normal rolls or hexagons.

Similar patterns have also been reported in the scientific literature, in particular in papers written in the early decades of the twentieth century. The worm-like patterns in thin layers of cream were called vermiculated rolls, and the toroidal cells in thicker layers, isolated cells. Both appear when there is a surface film on top of the convecting substance that hinders movement between the surface and the bulk of the liquid. In this case, the fatty cream is partially blocking the surface, so these patterns appear.

More recent convection research has tended to ignore these types of patterns, and the old experiments have often been considered as inaccurate because the fluids were impure; the different patterns were thought to be the result of impurity. We have tried to redress this. After seeing the patterns in Tia Maria, we carried out solutal convection experiments with simpler pure fluids that still show the same patterns. You can find a thorough account, both with Tia Maria and more conventional lab chemicals, in *Physica A* as detailed above.

JULYAN CARTWRIGHT, ORESTE PIRO, ANA VILLACAMPA

Honey Monster

How can an unopened jar of runny, clear honey suddenly begin to turn into a hardened block of sugar with no

obvious external stimulus? Jars that have remained clear for years can, over the space of a couple of weeks, change into solid sugar while the jar remains motionless on its shelf. Temperature does not seem to be a factor—the process can occur in winter or summer.

BILLY GILLIGAN

Beekeepers argue about this, as honeys from different sources behave differently. Honey is a supersaturated solution of various proportions of sugars (mainly glucose and fructose), and is full of insect scales, pollen grains, and organic molecules that encourage or interfere with crystalization. Glucose crystalizes readily, while fructose stubbornly stays in solution. Honeys like aloe honey, which is rich in glucose and nucleating particles, go grainy, while some kinds of eucalyptus honey stay sweet and liquid for years.

Unpredictably delayed crystalization means a nucleation center has formed by microbes, local drying, oxidation, or other chemical reactions. Crystalization can also be purely spontaneous, starting whenever enough molecules meet and form a seed crystal. Some sugars do this easily, others very rarely.

By seeding honey with crystals, or violently stirring air into it, you can force crystalization. Products made this way are sold as "creamed" honey. The syrup between the sludge crystals is runnier and less sweet than the original honey, because its sugar is locked into crystals. Gently warm some creamed honey in a microwave until it dissolves, compare the taste of the syrup with the sludge—you will be astonished.

JON RICHFIELD

I have seen this happen many times. The time before crystalization starts seems to depend on the source of the nectar the honey is made from. Oilseed rape honey will crystalize within a week or two of the bees making it. Heather honey never seems to crystalize. Fuchsia honey is extremely runny and, unlike any other I have seen, seems prone to fermentation, even when all the extracted honey comes from cells capped by the bees for storage. Even this crystalizes after a year or two.

PAT DONCASTER

Gurgle Time

Does liquid pouring from an inverted bottle flow faster at the beginning and end of its expulsion or when it reaches the "glug-glug" point somewhere in the middle? And, whatever the case, what is the explanation for the different velocities?

RANDY BARON

Water flowing from an inverted bottle has no free surface. So the water coming out has to be replaced by something else because liquids do not expand or contract very much when the pressure changes. In the case of a thin-walled plastic bottle, the volume can be replaced by the walls of the bottle being pushed in by air pressure as the water runs out, so this is what happens first.

Once this has happened (or in the case of a glass bottle, immediately) another replacement mechanism is needed and bubbles of air have to enter through the neck of the bottle. Essentially, the bubbles and the water escaping have to take turns, coming in and out, respectively, giving rise to the glug-glug effect.

Two other important factors affect the rate of flow. Firstly, if there is a significant amount of gas inside the bottle above the liquid, this can expand to replace the volume of water lost. This process goes on until the reduced pressure of

the gas is just sufficient to support the height of liquid above the exit. Then glug-glug starts again.

The second factor is swirl. If the bottle is handled so that the liquid swirls, the water moves to the outside of the neck and allows a column of air relatively free passage up the center. By actually moving the bottle in small circles before releasing the liquid, you can get a very effective tornado in a bottle.

These effects are important in industrial separating devices called hydrocyclones, which are shaped something like an inverted milk bottle. It is possible to tell how effectively these devices are working simply by viewing the pattern of discharge from the bottom, described as rope, cone, or spray.

MARTIN PITT
Department of Chemical and Process Engineering
University of Sheffield, UK

Note to bar staff in a hurry: A little experimentation in the New Scientist *laboratory confirms that whichever angle the bottle is held at, flow is fastest when the bottle is full, because this is when there is greatest pressure on the liquid at the mouth of the bottle.*

To pour out the contents of a bottle quickly, it is far more effective to keep it at an angle than simply turn it upside down. This is because the angled pour avoids the gurgling that slows the passage of liquid through the neck of the bottle. For an extra edge, follow Martin Pitt's advice: spin the bottle and then invert it while continuing to rotate it rapidly about its axis.

We found that a 750-milliliter wine bottle emptied in 9.9 seconds if inverted, but in 8.1 seconds if held at 45 degrees. Swirling the bottle so that a little tornado forms in its neck, allowing air to enter continuously and replace the liquid, brings pouring time down to just 7.7 seconds.

In all cases, the rate of emission slows as the head

of water above the neck falls. Dividing the volume of water in the bottle into equal thirds, the first, second, and third volumes left the inverted bottle in 2.5, 3.5, and 3.8 seconds; the angled bottle in 2.0, 2.4, and 3.7 seconds, and the swirling bottle in 2.0, 2.3, and 3.3 seconds. The swirling technique, although very smooth once perfected, is not recommended for high-speed pouring of beer or any drink containing gas.—Ed.

Changing Tastes

Monosodium glutamate is a common flavor enhancer that is used particularly in Chinese and Japanese cooking. Why is it so popular in these cuisines and, more pertinently, how does it enhance the flavor of food?

MICHAEL STUART

Monosodium glutamate, or MSG, is presumably most commonly used in oriental cooking for traditional reasons. For thousands of years the Japanese have incorporated a type of seaweed known as kombu in their cooking to make food taste better. It was not until 1908, however, that the actual ingredient in kombu responsible for improvement in flavor was identified as glutamate.

From then until 1956, glutamate was produced commercially in Japan by a very slow and expensive means of extraction. Then large-scale industrial production began and has continued, mainly involving the fermentation of natural substances such as molasses from sugar beet or sugar cane. Today, hundreds of thousands of tons of MSG are produced all over the world.

Monosodium glutamate contains 78.2 percent glutamate, 12.2 percent sodium, and 9.6 percent water. Glutamate, or free glutamic acid, is an amino acid that can be found naturally in protein-containing foods such as meat, vegetable, poultry, and milk. Roquefort and Parmesan cheese contain a lot of it. The glutamate in commercially produced MSG, however, is different from that found in plants and animals. Natural glutamate consists solely of L-glutamic acid, whereas the artificial variety contains L-glutamic acid plus D-glutamic acid, pyroglutamic acid, and other chemicals.

It is widely known that Chinese and Japanese food contains MSG, but people don't seem to be aware that it is also used in foods in other parts of the world. In Italy, for example, it is used in pizzas and lasagne; in the U.S. it is used in chowders and stews; and in Britain it can be found in snack foods such as potato crisps and cereals.

It is thought that MSG intensifies the naturally occurring "fifth taste" in some food—the other, better known, four tastes being sweet, sour, bitter, and salt. This fifth taste is known as umami in Japanese, and is often described as a savory, broth-like, or meaty taste.

Umami was first identified as a taste in 1908 by Kikunae Ikeda of the Tokyo Imperial University, at the same time that glutamate was discovered in kombu. It makes good evolutionary sense that we should have the ability to taste glutamate, because it is the most abundant amino acid found in natural foods.

John Prescott, associate professor at the Sensory Research Science Center at the University of Chicago, suggests that umami signals the presence of protein in food, just as sweetness indicates energy-giving carbohydrates, bitterness alerts us to toxins, saltiness to a need for minerals, and sourness to spoilage. A team of scientists has even identified a receptor for umami, which is a modified form of a molecule known as mGluR4.

MARK BOLLIE

Curious Cuppa

When you add a few drops of lemon juice to a cup of black tea, the color of the tea lightens considerably and very quickly. Why?

STUART ROBB

The simple answer to this question is that adding lemon juice alters the acidity of the tea and the color change is an indication of this, in the same way that litmus paper changes color.

A similar effect can be observed by substituting the tea with some cooked red cabbage juice.

ARON

Tea leaves are rich in a group of chemicals known as polyphenols that amazingly account for almost one-third of the weight of the dried leaf. Both the color of the tea and much of its taste are due to these compounds.

One group of polyphenols, the thearubigins, are the red-brown pigments found in black tea and constitute between 7 percent and 20 percent of the weight of dried black tea.

The color of black tea is also influenced by the concentration of hydrogen ions in the water. Thearubigins in tea are weakly ionizing acids and the anions (negatively charged ions) they produce are highly colored. If the water used to brew tea is alkaline, the color of the tea will be deeper due to greater ionization of the thearubigins.

If lemon juice, which is an acid, is added to the tea, the hydrogen ions suppress the ionization of thearubigins, and that makes the tea lighter.

Interestingly, the theaflavins—the yellow-colored polyphenols in black tea—are not involved in the change in color that is associated with a change in acidity.

JOHAN UYS

Indestructible Wine

I've just returned from a holiday in Madeira. I learned that old bottles of Madeira wine—a fortified wine similar to port and sherry—should be stored in an upright position. Bottles stored in this way are still drinkable centuries later. However, most other bottles of wine should be stored lying down to keep the cork moist and intact. Why is Madeira different—surely its cork will dry out too?

CRISTINA MARIANA

Old bottles of Madeira wine don't have to be stored in an upright position, but unlike other wines, it won't do them much harm.

Once wines have been bottled, oxygen becomes the enemy. It oxidizes the wine, resulting in an unpleasant odor and taste. The purpose of the cork is to keep out all oxygen except the small amount in the neck of the bottle. But because corks dry out and shrink, bottles stored upright will eventually let air in to oxidize the wine. Hence the typical advice to store wine bottles on their sides, keeping the cork moist.

Madeira wine, like sherry and port, is fortified by the addition of brandy before fermentation is complete. This means some residual sugar remains in the wine because the increased alcohol concentration kills off the yeast. Another result of this process is, of course, to make a more alcoholic wine (usually between 16 and 20 percent by volume instead of between 10 and 13 percent).

This increased alcohol and sugar content tends to protect fortified wine from oxidation, so the danger is lessened. However, some oxidation will still occur if oxygen is present.

Madeira, however, is a special case. It tastes better when it's somewhat oxidized, a characteristic that was accidentally discovered and then deliberately exploited in the eighteenth century by shipping barrels of it on sailing vessels on long journeys through tropical regions. Indeed, the term used for an oxidized dry wine is "maderized," obviously derived from "Madeira." Therefore, the risk of further maderization to a bottle of Madeira—from a dried-out cork, say—is not as serious as it would be with other wines.

Why, though, might upright storage be recommended? Between 5 and 10 percent of wine corks rot when kept wet, and bottles sealed with those corks will eventually acquire the moldy smell of rotten cork. Such a bottle is called "corked" when it is opened and sampled, hence the routine of smelling the cork before pouring the wine. If a bottle of Madeira is stored upright, the cork will never be wet and the bottle will never be corked. So if the risk of oxidation is considered a matter of little concern, and the risk of a moldy cork is a matter of greater concern, then the bottle should be stored upright. Of course, the very best solution would be to use only superior corks when bottling Madeira.

Will it still be drinkable centuries later? A couple of years ago I had the privilege of opening, decanting, and tasting about 2 ounces of an 1814 Madeira. It was still drinkable—not fine, but drinkable nonetheless. It had been recorked every 25 years or so. Its name was Violet, the same as my wife, so I kept the bottle. In those days, Madeiras were often labeled with the name of the ship in which they were transported to the U.S.

EDWARD HOBBS
Wine consultant
Wellesley College, Massachusetts

Vintage Madeira is quite capable of outlasting its cork. The practice therefore is to recork each bottle every few decades. A few shippers even list these recorking dates on their labels, in addition to the vintage date and the grape type. The oxidized state of the wine allows the process to be carried out with a fair degree of confidence, whereas the same process applied to port, sherry, or an unfortified wine would risk spoiling the contents.

The process by which Madeira is deliberately allowed to oxidize, known as estufagem, was discovered by accident after barrels of wine that had been sent on the long journey across the tropics to the New World were found to take on a pleasant color and taste.

For centuries, producers continued to send out their Madeira in barrels to act as ballast for ships and to improve its flavor. Now the barrels are simply kept at tropical temperatures of up to 120°F for about three months in the island lofts of the wine shippers.

MARK MCKEGROW

A Long Drink

What is the maximum length of a vertical straw with which you can drink cola?

BHARGAV

If you applied an absolute vacuum above a nonvolatile liquid, then the maximum height you could suck it up a vertical pipe would be reached when the hydrostatic head pressure of the column of liquid equals one atmosphere (101,325 pascals). This pressure is given by $p \times g \times h$, where p is the fluid density, g is the gravitational acceleration (32 feet per second per second) and h is height. For water, which has a density of 62.4 pounds per cubic foot, this gives a maximum height of about 34 feet.

However, because water has a vapor pressure of 3,536 pascals (at 80 °F), it will begin to boil before you reach a perfect vacuum. So the maximum vacuum pressure that you could apply is 101,325-3536 = 97,789 pascals. This gives a maximum height of 32.7 feet.

In the case of a soft drink, things are more complicated, because the dissolved carbon dioxide will start to "boil" out of solution under vacuum. If you sucked extremely slowly, first of all you would only get CO_2 and then, when you had removed the gas, you would get flat soft drink. If you sucked very quickly, then you might get the drink to rise up before the CO_2 nucleated and formed bubbles. More likely you would get a froth of liquid and CO_2 bubbles, and you might actually be able to suck this up to a much greater height because the effective density of the foamy mixture would be lower than pure liquid water. At intermediate suction rates, the foam bubbles would coalesce and you would be limited to a lower column height.

The exact answer depends on how much dissolved CO_2 you want left in your drink and the maximum rate at which you can suck. You would also need more than an ordinary drinking straw, because plastic ones collapse under moderate vacuum pressures.

Simon Iveson
Department of Chemical Engineering
University of Newcastle, New South Wales, Australia

Using a very long, thick-walled, plastic pressure tube, 15-year-old pupils can usually manage a six-foot suck to get a drink. By sucking, sealing the tube with the tongue, breathing and sucking repeatedly, a 14-foot lift is easily achieved. This is the highest my pupils have managed because the next option is to stand on a step ladder at the top of a stairwell and this not a good idea when you have a class of 30.

I suppose that this height is approaching the limit. The pressure reduction in the mouth is the same as that in the top of the tube, so it becomes difficult to suck against the external pressure, and it also becomes difficult to get your tongue out of the end of the tube.

There is also the problem of internal pressure in the lungs which, if the throat is opened and air is expelled into the tube vacuum, can fall dramatically. It is wise to stop before this point.

Keith Sherratt

Be careful experimenting with the limits of your ability to suck up fluids. Not only could you choke unpleasantly, but strong suction can cause blood blisters in your mouth.

In the Kalahari, as recently as a decade or two ago, the !Kung people sometimes had to suck water out of narrow holes in rocks. In dry seasons, they would join reeds into long straws and the men who were able to suck it up far enough would spit it into a communal container.

Jon Richfield

Shock Value

Could someone please tell me why and how fabric softeners reduce the amount of static electricity in clothes?

Johanna

Static electricity is an imbalance of electric charge: a lack or overabundance of electrons on the surface of the material. This typically occurs by "tribocharging" when two materials are brought into contact then separated; electrons are exchanged by the materials, leaving one with a positive charge and the other with a negative charge. Friction between the two materials can enhance this charge-separation process.

Under normal atmospheric conditions, fibers such as cotton and wool have a relatively high moisture content, which makes them slightly conductive. This prevents the charge separation from occurring by allowing static electricity to be conducted away. However, synthetic materials have a high surface electrical resistance particularly when humidity is low—and this prevents the charge dissipating. A layer of fabric softener simply reduces the electrical resistance of the surface of fabrics.

Paul Thompson

Static build-up in clothing is caused by fiber-to-fiber, fiber-to-person and even fiber-to-air friction, and depends on the type of fiber from which the garment is made. The amount of static build-up is also highly dependent on the relative humidity—the higher the humidity the lower the charge. Fibers such as rayon, silk, wool, cotton, and linen have high moisture "regain"—their fibers absorb a great deal of moisture at a given humidity from a bone dry condition—and are low in static. Fibers such as polyester, acrylic, and polypropylene, having low moisture regain, are high in static.

Antistatic finishes or sprays come in two types. The first are made of molecules that contain polar groups, in which charge is unevenly distributed, and these act as conductors to dissipate static charge. The second type are humectants, or water-absorbing materials, that also permit the textile to dissipate static electricity. The increased moisture present on the surface or within the fiber itself increases electrical conductance, helping to drain away the charge.

Textile technologists can design fibers and fabrics to minimize static electricity. In carpets, a small percentage of fibers (up to 3 percent) can have either a carbon core or a carbon strip to drain away static charge. Carpets and upholstery fabrics may also be made with carbon lampblack mixed into the latex or hot-melt backing material for the same purpose.

If the carpet is made of yarn spun from staple fibers, a small percentage of stainless steel fibers or fibers coated with aluminum or silver vapor may be incorporated into the blend to reduce static electricity. However, less than 5 percent of this type of fiber can be used because otherwise the fabric takes on a gray tinge.

BOB WAGNER

Fabric softeners contain a type of compound called a surface active agent or surfactant. It's a cationic surfactant, meaning it's a long molecule (rather like an oil or a fat) with a positive charge at one end. Often the surfactant used

is an ammonium compound in which the nitrogen atom is surrounded by four organic groups.

During the washing process the negative charge that forms on the surface of the fabric draws the positive end of the surfactant molecules to itself. These long oily molecules then lubricate the fibers to prevent the friction that causes static cling. It makes ironing easier, and allows the weave to relax and supply that soft, fluffy feel.

RICHARD PHILLIPS

Honey, I'm Bendy

Why does a slice of bread spread with honey gradually become concave?

DONAL TROLLOPE

My wife has assured me that her bread doesn't have time to go concave when spread with honey. However, for those folk who chomp their honeyed bread in a more leisurely fashion, there is a simple explanation.

Bread is approximately 40 percent water while honey is a strong solution containing approximately 80 percent sugar. This means that moisture is drawn out of the bread and into the honey by osmosis. Removing the water makes the bread shrink, but only on the side exposed to the honey. This causes the bread to become concave.

This is less likely to happen, of course, if you butter your bread before spreading the honey. Butter forms a water-impermeable layer that protects the bread from dehydration by the honey.

PETER BURSZTYN

Gray Matter

The surfaces of the incandescent light bulbs where I work become progressively grayer over time. Why?

KIRSTY RHODE

The graying of the inner surfaces of incandescent bulbs is the result of gradual evaporation of tungsten from the filament while the light is on. This evaporation eventually makes the filament so thin it burns out.

Various methods have been developed to reduce graying. Filaments of the first incandescent lamps burnt in a vacuum, but it was soon found that introducing inert gas to the bulb reduced the rate of graying. A mixture of nitrogen and argon is used today. In addition, "getters," reactive metals such as tantalum and titanium, can be placed near the filament to attract the tungsten so that it is not deposited on the glass. Alternatively, a small amount of abrasive tungsten powder can be placed in the bulb. Shaking it occasionally will remove the gray coating from the surface of the glass.

Graying can be almost eliminated by introducing a small amount of the halogens iodine and bromine. As tungsten evaporates from the filament, it reacts with the halogens which then redeposit the tungsten on the filament. This keeps the bulb wall clean. To prevent the tungsten halides from condensing on the bulb and breaking the cycle, the temperature of the bulb wall must be at least 900°F. This is too hot for glass bulbs, which normally operate at about 300°F, so fused quartz (silicon dioxide) must be used instead.

Compared with ordinary incandescent lamps, quartz-halogen lamps have longer lives and maintain their light output over time. For example, a quartz-halogen lamp with a 2,000-hour life will have dimmed by less than 5 percent by the time it burns out. When an incandescent lamp with

a 1,000-hour life burns out, it will have dimmed by more than 15 percent.

Ross Firestone

This can be explained by the fact that lights work not by emitting light but by sucking dark. "Dark sucker" theory is too complex to be described here in detail, but it proves the existence of dark, that dark is heavier than light, that dark is colored, and that it travels faster than light.

To answer your question, a bulb becomes darker over time because of all the dark it has sucked in. Similarly, a candle, which is a primitive type of dark sucker, has a white wick when new and this becomes black when used, due to all the dark which has been sucked into it.

Ken Walke

Readers should note that the revolutionary "dark sucker" theory has yet to win widespread support from the scientific community.—Ed.

Heated Hop

I was sitting with a group of friends in the bar last night and, as we contemplated our pints of ale, we wondered: why does beer go flat when it gets warm? Not only that, but the effect seems to be more pronounced with lager than with ale.

Jon Shaw

The answer lies in the behavior of gases, and their solubility in water. Most beers are dilute solutions of sugars, gases, organic acids, and other complex compounds, and (hopefully) alcohols.

The gas which gives fizzy drinks their distinctive fizz is carbon dioxide. In the case of British-style real ales, the CO_2 is generated by the action of yeasts on residual sugars in the drink, whereas in most beers, including British-made lagers, the conditioning gas is added artificially at the brewery or at the point of sale.

The problem arises because the solubility of CO_2 is related to the temperature of the solvent in which it is dissolved, in this case the beer. More gas can dissolve in cold beer than in warm beer. This is also why fish such as trout and salmon, which need a lot of oxygen, live in cold mountain streams and rivers, because the amount of oxygen dissolved in these environments is so much higher.

When the beer is served from the pump it will contain a certain concentration of dissolved CO_2, but, as the drink heats up under the influence of a sweaty hand in a warm room, its ability to hold its CO_2 in solution decreases.

This excess gas is then released into the atmosphere through the bubbles that you see rising in the beer, and the drink consequently goes flat. Other volatile compounds from the malt and the hops vaporize faster and you may notice that the beer also smells different.

The difference your correspondent observes between lager and bitter is mainly caused by two factors. The first is that lagers are generally served much colder than bitters, to disguise the fact that they have less taste (this is because they contain fewer fruity esters and longer-chain alcohols, as a result of the lower fermentation temperatures and different yeast strains that are employed in their manufacture). The bigger temperature difference between the beer and the air means lagers warm up faster than bitters. Their rate of loss of CO_2 is consequently much higher, and they go flat faster.

Secondly, lagers are generally more carbonated than bitters, so are fizzier at the point of purchase and therefore have more CO_2 to lose in the first place. This extra carbonation, like the lower serving temperature, is usually used to

camouflage the lack of flavor that is found in most British-brewed lagers.

The answer to the problem, of course, is to drink your beer faster, or sit in cooler pubs.

GEOFF NICHOLSON

4 Our Universe

Planet Pinball

If all the matter in the universe was created in the big bang, and if the universe has been expanding since then, by what mechanism can two old galaxies collide?

DON JEWETT

Galaxies can collide because the expansion of the universe is the expansion of space itself, not the movement of matter through space. Local movement is independent of this overall expansion. The Andromeda galaxy is actually moving towards us.

GRANT THOMPSON

The big bang was not like a normal explosion, in which fragments of a lump of matter are blown out. Rather, the big bang set space itself expanding.

A common cosmological analogy is to think of galaxies as paper dots on the surface of a balloon. As the balloon inflates, the galaxies move apart, because the very space between them grows. In this analogy, it is the surface of the balloon, not the volume within, that represents the three-dimensional universe.

Galaxies may have their own trajectories across the surface of the balloon, pulled about by the gravity of other galaxies. This local movement is distinct from the expansion of space itself and means galaxies can collide.

BOGDAN KAMENICKY

The universe overall is expanding. But gravity makes sure not all matter is traveling away from the center. Look at the Earth as it orbits the sun. Half our time is spent traveling away from the center of the universe (wherever or whatever that is), and the other half traveling back.

STEVE MINEAR
Emory University
Atlanta, Georgia

Which Way to Turn?

If you took a compass into space, how far above the Earth's surface would you have to go before it stopped showing which way is north? Further into space it would presumably respond to the magnetic field of the sun or planets, but how could we interpret its reading?

BEN

The magnetic field of the Earth looks like a dipole (the shape formed by iron filings around a bar magnet), although the Earth's is rotated about the field's axis to form a three-dimensional shape. This extends to about 40,000 miles into space. On the ground we use a compass in two dimensions. In space you can use a 3D "compass" to map out the Earth's magnetic field, again giving an indication of north.

Beyond 40,000 miles into space, in the direction of the sun, we exit the Earth's magnetosphere and pass into the solar wind, which also carries the sun's magnetic field. During undisturbed solar periods the sun's field is shaped like a spiral, thanks to the sun's rotation, in the same way that a hosepipe whirled over your head emits a spiral of water.

Magnetic field measurements are made by interplanetary spacecraft to understand how the sun's magnetic field and solar wind interact with the Earth's magnetic field. For instance, auroral displays are generated by the solar and terrestrial magnetic fields interlinking to allow solar wind plasma to enter the atmosphere.

On the opposite side of the Earth from the sun, the Earth's magnetic field is pulled into a long magnetic tail by its interaction with the solar wind, typically to 4 million miles or more. A compass in this geomagnetic tail would point along the tail, either towards or away from the Earth.

It is interesting to note that if we left the solar system beyond the "heliopause" where the solar wind ceases to have an effect and traveled into interstellar space (approximately 150 astronomical units from Earth), our compass would start to measure the galactic field. Here, our magnetic field measurements might point towards the constellation Pyxis, appropriately better known as the Compass.

STEVE MILAN

Turn Left at Mars

As a hiker and a pilot, if I wish to travel from point A to point B, I can follow a magnetic heading on my compass to get there. What mechanism is used by astronauts or automatic probes in space travel to ensure that spacecraft are going the right way?

HOWARD ARBER

Navigation requires that you know where you are relative to your destination, and how to negotiate the route. To do this in space, knowing your attitude is as crucial as knowing your position, so the first thing to do is find and track the sun and a known, conspicuous, distant star.

Sirius is a good example, but it lies relatively near the celestial equator, so sometimes the sun gets in the way. Canopus is better: almost as bright, and far south in the celestial

sphere, well away from the sun. From the positions of such stars and the sun, you can calculate your attitude, and can locate other bodies by radar, data from mission control, or visual observation. Gyroscopes can then damp oscillations and detect minute changes in attitude, while Doppler measurements let you calculate your velocity.

In space, knowing your trajectory relative to major masses in the solar system permits you the luxury of navigating by dead reckoning for millions of miles. Only when you apply thrust to adjust your course, get close enough to bump into things, or need to maneuver into a precise orbit is it necessary to check on your exact situation and make corrections.

JON RICHFIELD

The Apollo missions depended on ground-based radar, which could determine their position and range, plus, using Doppler measurements, their radial velocity. Course changes were computed on the ground and radioed to the crew. The figures were then punched into the on-board computer, which took care of the engine burns. As a back-up, the crews were trained to be able to take star readings and calculate course changes for themselves. This was never required, although for one course adjustment on Apollo 13, the computer was not available and the burn had to be controlled manually.

ALEX SWANSON

No More Moon

What would be the effect on the Earth if an alien spaceship came along and dragged the moon away?

STEVEN NAIRN

Any alien spaceship stealing the moon would unleash a devastating chain of events that would ultimately spell the end for life on Earth.

The most immediate difference would be the disappearance of the tides. Both the sun and moon influence the tides on Earth, but the moon is the dominant force. Remove the moon and the daily rush of the tides would recede to a gentle ripple.

The next omen of doom would be wild swings in the Earth's rotational axis from a position almost perpendicular to the ecliptic plane all the way to being practically parallel to it. These swings would provoke drastic climate changes: when the axis points straight up, each point on the globe would receive a constant amount of heat throughout the year but, when the axis lies parallel to the ecliptic, Earthlings would spend six months of the year sweltering under the unending blaze of the sun, only to spin round and shiver for the next six months, hidden on the frigid surface of the Earth's dark side.

Of all calamities, though, the creature to be pitied first is the marine organism called "nautilus." This mollusc lives in an elegant shell shaped like a perfect spiral partitioned off into compartments. The nautilus only lives in the outermost partition, and each day adds a new layer to its shell. At the end of each month, when the moon has completed one revolution around Earth, the nautilus abandons its current compartment, closes it up with a partition, and moves into a new one. Scientists have proved that the number of layers making up a chamber are directly linked to the number of days it takes the moon to circle the Earth. Remove the moon and the nautilus lies stranded, forever locked in the same chamber and wishing ruefully for the days when it could look forward to a new home.

ANDREW TURPIN

The moon and the Earth both have a gravitational effect on each other. They orbit about a point in between them and, as a pair, in turn they rotate around the sun. If the moon was suddenly taken away by aliens, the pull from the moon would disappear, thus unbalancing Earth's orbit. This would lead to the Earth plunging out of its current orbit in a direction which depends on the position of the moon and Earth at that time. It would probably result in a more elliptical orbit and greater extremes in temperature and massive climate changes would make our planet uninhabitable.

Knowing this, we should all worship the moon because we might have not evolved into what we are without it.

Hovick Boughosyan

The end of tides would have a major detrimental effect on coastal ecosystems. Mangroves, for example, rely on regular tidal motions to sweep in nutrients and such. It would also change the patterns of ocean currents, causing major climate change.

Additionally, a major source of nighttime light would disappear. This would affect the behavior of all nocturnal animals and the synchronization of behavior associated with the lunar period. Owls would find it more difficult to hunt and insects would find it harder to find a mate because they fly up towards the moon.

Simon Iveson

To reassure the many nervous readers who contacted us, we can confirm that we have no inside knowledge of an alien plot to remove the moon. It would seem an unlikely stunt, even for a civilization with a sense of humor that is advanced many millions of years beyond our own.—Ed.

Low-Gravity Lager

Apparently NASA is aiming to brew a beer in space. The yeast can neither sink nor rise as in traditional beer production and any carbon dioxide that is produced will not rise to the surface. So how will the beer ferment and will the final product be anything like Earth beer?

ROGER CRYNE

NASA is indeed interested in the questions raised by brewing beer in a microgravity environment. Scientists who study the physics of gas-liquid mixtures would like to understand, for example, what happens when there is no buoyancy to bring the bubbles to the surface of a fizzy liquid, and the characteristics of fermentation in microgravity.

Two separate space shuttle experiments tackled these questions. The first investigated how well yeast performs in orbital free fall—not only to see if brewing space-beer might be possible, but also to provide valuable information to pharmaceutical companies with a keen interest in the biology of orbiting microbes.

The space-beer turned out essentially the same as that brewed on Earth: its specific gravity and the yeast's performance when used to brew subsequent batches of beer was comparable to that of control samples on Earth. However, the total yeast cell count and the percentage of live cells in the space sample were lower. Despite this, the fermentation was significantly more efficient. This raises the question of whether we can modify the fermentation process, or the yeast itself, to reproduce this effect on Earth.

The second experiment, flown on the shuttle by the Coca-Cola Company, was to test its system for dispensing Coke in a weightless environment. The challenge was to dispense a fizzy beverage yet keep the gas in solution until the cola is drunk. Because bubbles don't rise in free-fall, changes in temperature, pressure, or even physical agitation

tend to cause the whole thing to degenerate into a foamy mess.

A computer-controlled device adjusted the temperature of the drink during mixing and dispensing, and minimized agitation by dispensing the drink into a collapsible bag inside a pressurized bottle. The pressure around the bag was slowly released as it filled with drink, keeping the drink under constant pressure and preventing the gas from coming out of solution too quickly. The end result was a space version of the world-famous fizzy drink.

DANIEL SMITH

Gnab Gib

If antimatter had prevailed over matter after the big bang, would anything be different in our universe?

SAM HOPKINS

Living in a universe where antimatter predominates would be akin to living in our mirror image. We wouldn't actually notice anything different. All the positive charges would be negative and vice versa, but to someone studying physics in an antimatter universe the idea of positrons orbiting negatively charged nuclei would be as natural as electrons orbiting positively charged nuclei is in ours.

MIKE FOLLOWS

Until the early 1960s the answer would have been: "As far as we can tell, there is no reason it should." Since then certain subtle differences between matter and antimatter have emerged and the answer is: "Probably, but it is too early to say how and why." We don't even know yet why matter prevailed in our universe. It is most plausibly because of some random asymmetry, but it is also possible that antimatter was unstable in the circumstances that reigned at that point.

So has an antimatter counterpart to our universe existed, with life and so on? We don't even know that for sure. The fact that stellar and biological evolution are so slow in our matter universe does not mean that no faster universe is possible. Even today, some people suggest that in the super-fast reactions in the quark soup of neutron stars, living structures with the complexity of civilizations might arise and pass in what to us would be the blink of an eye. Maybe by the time antimatter had vanished, it had generated worlds of life as complex as our own. Or, because antimatter in some ways is seen as moving backwards in time, perhaps it has not yet started.

JON RICHFIELD

What makes your correspondent so sure antimatter didn't prevail?

VILNIS VESMA

5 Our Planet

Dump It in the Mantle

Would encasing nuclear waste in concrete and then burying it in a tectonic subduction zone for the Earth's mantle to consume be an effective way of disposing of it? If not, why not?

ALEC PAPPAS

Subduction zone insertion was one idea proposed for the disposal of radioactive waste during the early history of atomic energy. Other ideas included a serious proposal to dump canisters of waste on the Antarctic snow and leave them to melt their way to the bottom of the ice sheet.

In fact, subduction zone insertion is perfectly sound in theory, but there are significant practical problems. The zones are inherently unstable and unpredictable, and the sediment on top of the subducting ocean crust plate tends to get scraped off rather than being carried into the mantle, to form what is known as an accretionary prism. This could lead to the waste being squeezed back to the seabed in the future. Drilling it deep into the basalt of the crust may solve this, but at the depths typically encountered in subduction zones, drilling is all but impossible.

Some have proposed a more elegant seabed solution, which is to insert the waste into the deep clays that cover most of the shallower, geologically stable abyssal plains. This could be achieved either by drilling holes and slotting waste canisters into them, or by dropping waste through the ocean

in rocket-shaped "penetrometers," which would use the kinetic energy of their descent to burrow many tens of feet into the soft clay. Though not without their problems, these methods have the advantage of inserting the waste into a stable, impermeable environment where any nasty leaks from the canister would be effectively absorbed by clay particles in the surrounding ooze.

SAM LITTLE

Water, Water . . .

How can oceanographers tell the average depth of the ocean? In calculations of sea levels, currently considered a matter of great importance, they can seemingly tell to the nearest inch if the water is rising or falling. Surely even on a windless day the ocean surface rises and falls by more than a few inches, simply due to local wave action, not to mention the changes caused by tides and swells.

ROGER SHARP

The ocean floor is mapped using satellite measurements of gravity anomalies. The ocean floor is static, so the measurements can be averaged over a long period to give quite a high level of precision—though unfortunately not to the nearest inch. In any case, knowing the average depth of the ocean is not really important for measuring sea-level rise.

Sea level is measured at the coast by tide gauges, and globally by satellite altimeters. The gauge can be an old-fashioned float type, or an acoustic sensor or radar. The rapid changes in height caused by waves can be smoothed out by mounting the tide gauge in a "stilling well," essentially a vertical tube with an opening at the bottom that is about one-tenth the diameter of the tube itself. The stilling well damps out the waves but allows tides and surges to be measured.

With satellite altimeters the picture is more complicated, as many adjustments have to be made to account for air pressure, water vapor content, and the like in the atmosphere, as well as scattering by waves and the effects of tides. The processing is very complex and even then the altimeters need to be calibrated with tide gauges.

For both tide gauges and satellite altimeters, the key to reaching millimeter accuracy is averaging. Satellites average over an area about 4 miles in diameter, and data from both satellites and tide gauges is averaged over time. In this way tides, waves, storms, and even the seasonal cycle are removed from the data, allowing mean sea level to be determined very accurately.

The global database of monthly mean tide gauge records is kept in the UK by the Permanent Service for Mean Sea Level. This data shows that over the past century sea level rose at about 2 millimeters per year. Over the past decade, the rate of sea level rise has been nearer 3 millimeters per year, but it is too soon to tell if this is a temporary fluctuation or a long-term change.

Simon Holgate
Permanent Service for Mean Sea Level
Proudman Oceanographic Laboratory,
Bidston Observatory, Prenton, Merseyside, UK

Sea level is measured using tide gauges, which are kept sheltered from the waves by a stilling well. These instruments can only measure the relative height of land and sea, and it is worth remembering that not only the sea changes height: the land moves too as a result of plate tectonics and other natural processes. Also, tide gauges at major ports are affected by urbanization, as the weight of large cities can cause local subsidence, which may result in a spurious sea-level rise. Intensive settlement has accelerated the removal of groundwater beneath cities such as Adelaide in South Australia, causing them to sink relative to the sea. All these

factors make satellite altimetry a more accurate technique for measuring sea-level rise.

Against this background it seems almost miraculous that sea level has been measured as rising at between 1.7 to 2.4 millimeters a year, as a result of global warming. This is mainly due to the thermal expansion of water, with a small contribution from the melting of ice on land.

MIKE FOLLOWS

Radar altimeters on satellites are used to measure sea level. If the altitude of the satellite is known, the height of the ocean can be determined to within a few inches.

The average of many orbits over a particular location can be used to map short-term changes in sea level induced by variations such as seasonal heating, river discharges, and evaporation. Over a longer period, averaging begins to show changes in sea level in the oceans caused by such phenomena as the Southern Oscillation and the North Atlantic Oscillation. Averaging of all orbits over all oceans over time yields a measurement of global sea level that shows that it is rising by about 2.3 millimeters a year.

One of the most valuable aspects of the resulting map of sea level is that it mirrors features on the seabed. Extra mass in the form of seamounts increases gravity and slightly raises the ocean surface above the feature. A depression in the seabed lowers sea level locally.

Satellite altimeters have taken only a few years to map previously uncharted ocean bottom that would have taken 100 years to map using conventional techniques. The information this yields is priceless. For example, the altimeter maps have been used to identify an impact crater south of New Zealand 12 miles in diameter. Detection of this crater using ship-borne techniques would have been virtually impossible and too expensive.

TED BRYANT
Associate Dean of Science
University of Wollongong, New South Wales, Australia

Hidden Depths

I always believed that the sea looked blue because it reflected the color of the sky. On vacation in Malta the sea was a very clear, deep azure blue inside caves where there was no reflected sky. What caused this color?

PETER SCOTT

Seawater appears blue because it is a very good absorber of all wavelengths of light, except for the shorter blue wavelengths, which are scattered effectively. The light attenuation is caused by the combined absorption and scattering properties of everything in the water, along with the water itself.

Changes in the sea's color are primarily due to changes in the type and concentration of plankton. Tropical oceans are clear because they are lacking in suspended sediment and plankton, which contrasts with the popular misconception that tropical waters have a high biological productivity. In fact, they are virtually sterile compared with the cooler, plankton-rich temperate ocean regions. Inorganic particulates and dissolved matter also reflect and absorb light, which affects the clarity of the water.

JOHAN UYS

The effect is caused by the selective absorption of light by water molecules, chiefly the oxygen component, that take out the red end of the visible light spectrum. In a similar way, ice masses at the poles and big icebergs look blue.

ALBERT DAY

Reflection of light contributes to the color of the open sea, but does not determine it. Even pure water is slightly blue-green, because it filters out the red and orange content of light. However, impurities in seawater, especially organic substances, affect its appearance far more drastically.

In caves like those described, the light coming in must travel through a greater thickness of seawater than the light we usually see. The strong absorption of wavelengths other than blue and green intensifies the ethereal effect. In fact such light contains so little red that navy personnel who have been on submarine duty for several days find everything looks unnaturally ruddy when they return to the surface.

JON RICHFIELD

Blue Lake near Mount Gambier in South Australia is always blue, sun or no sun. The lake is situated in a limestone area and is saturated with calcium carbonate. The color comes from the greater scattering of blue light by very fine particles of the compound suspended in the water.

Seawater is normally supersaturated with calcium carbonate, but magnesium in the water tends to stop it precipitating out. However, this can occur when seawater comes into contact with the calcium carbonate mineral calcite in rock or soil. It is possible this what is happening in the caves of Malta.

ROBERT GERRITSE

Seasonal Shift

I was always under the impression that the equinoxes fell on March 21 and September 21, dividing the year into four equal parts along with the solstices. However, I often read that the equinox will fall on a day other than the 21st. Surely there has to be an equal division of the seasons, relying on the Earth's orbit around the sun? What could possibly change this?

KINGSLY RICHARD

The spring and autumn equinoxes are defined as the point in time when the sun is overhead at midday local time on the equator (in astronomical terms, the time at which the sun crosses the celestial equator). On the equinoxes there is an equal length of day and night everywhere in the world. The precise date of the equinoxes varies slightly; in the northern hemisphere the spring equinox usually falls on either March 20 or 21 and the autumn equinox on either September 22 or 23 (in the southern hemisphere the dates are reversed). This variation is simply because some years are leap years, so there is a shift in the calendar of a day or so relative to the seasons.

The equinoxes occur on exactly opposite sides of the Earth's orbit around the sun, but it is interesting that the dates on which they fall do not divide the year into two equal halves. Take the average dates of the equinoxes and the mean length of the year, and the autumn equinox falls 186 days after the spring equinox, whereas the spring equinox is only 179.25 days after the autumn equinox. This is because the Earth's orbit is elliptical and the Earth is closest to the sun in early January. In accordance with Kepler's second law, which states that a line joining a planet and the sun sweeps out equal areas in equal intervals of time, this is the part of the year when the angular velocity of the Earth in its orbit is greatest. As a result, the half of Earth's orbit from the autumn to the spring equinox takes less time to complete than the half between the spring and autumn equinox, when the Earth is further from the sun and moving more slowly. Consequently, spring and summer, during which there are more than 12 hours of daylight, last nearly seven days longer in the northern hemisphere than in the southern.

ROBERT HARVEY

The assumption that "there has to be an equal division of the seasons" is incorrect. Ptolemy, who flourished in A.D. 140, tried to account for the unequal lengths of the seasons. As I

describe in my book *Astronomy: From the Earth to the Universe* (www.solarcorona.net), current values show that spring lasts for 92 days, 19 hours; summer 93 days, 15 hours; autumn 89 days, 20 hours; and winter 89 days, 0 hours. Ptolemy deduced that the sun's orbit around the Earth, assumed to be circular for what were then thought to be obvious reasons, was not centered on the Earth, or that the large circle of the orbit had an epicycle.

Our current explanation, thanks to Johannes Kepler in 1609, is that the Earth's orbit around the sun is elliptical, with the speed of the Earth in its orbit varying according to Kepler's second law. The Earth is closest to the sun (the perihelion) in early January, so it is moving fastest in autumn and winter, accounting for those seasons being the shortest.

JAY PASACHOFF
Williams College, Massachusetts

Lava Wave

If I had to save myself by surfing down a molten lava flow, what would I be able to stand on that wouldn't melt from the heat of the lava?

BEN WILLIAMS (AGE 6)

Just take an old surfboard, punch lots of holes through it, and connect them to a water tank placed on top of the board. Water escaping through the holes will create the same effect that you can observe when spitting on a hot iron plate: the droplets dance on the plate for quite a long time because they are separated from the plate by a thin layer of steam, which is a bad heat conductor.

This effect would allow you to surf on the lava wave, because the board would be cushioned from the lava by the steam layer. The friction between the board and the lava would be virtually zero. I think that this creation should be known internationally as the sizzleboard!

RADKO ISTENIC

A lava surfboard must not only be melt-proof but it should also be less dense than the lava underneath and it should provide insulation for your feet.

If you are trapped at the summit of a volcano and you need to surf in order to escape, then you must use indigenous materials. Fortunately, volcanoes not only produce lava but also solid materials that are of roughly the same geological composition as lava but are less dense and more insulating because they contain gas bubbles.

A slab of this material, say 20 inches thick by 3 feet wide and 6 feet long, would float on molten lava and, just as important, it would melt quite slowly. I suspect that you could travel a mile or more before you would have to abandon it. Hopefully, by then, you would have been able to negotiate your way to an area of dry, cool ground.

However, if you know in advance that you will need to float on molten lava, you would be better off making a boat from a heat-resistant, or refractory, material that would not melt and would last as long as you needed it. The vertical sides of the boat would also protect you from the heat radiated by the lava far better than a surfboard would.

The temperature of molten lava is usually about 2,550°F although it can be as high as 3,000°F depending on its chemical composition, so the best material to use for a boat is high-purity alumina insulating refractory concrete. This is made of aluminum oxide, which melts at 3,600°F, hardly reacts with molten lava and contains hollow bubbles, so it is slightly less dense than the molten lava it will have to float on, and is a good insulator too.

To construct your boat, make a mold by digging a pit in the ground of the shape you want the outside of the boat to be, then pound the soil in the pit until it is compact and smooth. You should then line the mold with plastic sheeting

and mix the dry concrete with just enough water to form a stiff paste. Cover the plastic with a 4-inch layer of paste, line the concrete with another plastic sheet, and fill the remaining cavity with water to press the concrete into place while it sets. The boat will be ready after a week.

ROSS FIRESTONE

If the only consideration is melting point, your correspondent would not have much difficulty. Different types of lava melt at different temperatures, rhyolite at up to 1,650°F, dacite at up to 2,000°F, andesite at up to 2,200°F and basalt at up to 2,300°F. Steel, with a melting point of 2,550°F, would be fine, but to be really safe, how about tungsten, with a melting point of 6,191°F?

However, your feet would get a bit hot, so it would be better to use a nonmetallic insulating substance such as the following ceramics: Cr_2O_3 melts at about 4,100°F and even Al_2O_3 at about 3,720°F would be safe, and both would provide insulation for your feet.

I suspect, though, that the 6-year-old questioner might have difficulty getting hold of such materials, so I suggest using oak. All woods, and especially oak, form a protective carbonized layer when burnt, which slows further combustion. Indeed, when designing a timber structure one can allow for this layer to provide fireproofing. Timber structures are always designed a little over the size that is actually required so that their structural integrity is retained in case of fire. A thin steel plate lining the outside of the board would protect against abrasion if the surfer wanted to go back to the top to repeat the experiment.

MALCOLM NICKOLLS

The problem is not just keeping the heat away. When I was in the Sahara I saw those lovely sand dunes and thought that surfing down one of them would be fun. Sadly sand and wood do not have the right coefficient of sliding friction, and you don't so much surf as sit, even on the steepest dunes.

You can calculate coefficients of sliding friction using tables, and you should be looking for the sort of figures typical of waxed skis on snow. My tables are a little old, and there are no figures available for lava, so maybe one of your correspondent's first jobs if he chooses to become a volcanologist would be to calculate these coefficients of friction experimentally for different materials, so that a suitable escape craft can be designed.

Of course, the designer would have to be a quick and accurate worker or their career may not be quite as long as they might hope.

PETER BROOKS

Coast to Coast

If the oceans on Earth receded and eventually disappeared, there would be no coastline left. Likewise, if the seas kept rising, the total global coastline would reach zero as the waters lapped over Mount Everest. At some height in between there must be a maximum total coastline on the planet. Does anyone know where this would be in relation to today's sea level, and are we anywhere near such ideal conditions for a beach holiday?

BEN CADORET

The calculation is not so simple. As the letters below show, if we regard land masses as simple cones then the coastline initially grows longer as sea level falls. If we accept that actual topography is more complicated, then the answer becomes less clear. And finally, if we measure coastlines at progressively smaller scales, then we already have an infinite beach!—Ed.

Much more than half of the Earth's surface (some 71 percent) is covered by oceans and seas, so all land masses, however large, can be considered as islands. Islands are characterized by having their terrain sloping up from the coast, like a partially submerged cone. This means that a rising sea level will result in a shortening of the coastline and, conversely, a drop in sea level will result in a lengthening.

So the world's total coastline of some 534,000 miles can only become longer with a drop in sea level. The lengthening of coastline will continue until a point when seas and oceans make up less than half of the world's surface. These bodies of water can then be considered as lakes. Lakes are the inverse of islands and the length of coastline will decrease with each drop in water level.

Given that the sea-to-land ratio is nowhere near 50:50 and the average depth of the oceans is larger than the average height of land masses, the sea level that yields the longest coast is likely to be much lower than today's levels. Perhaps by a couple of miles.

To establish the point at which coastline reaches its maximum length, you would need to build a complex computer model of all land-mass and seabed topography, which could then be used to measure the length of coastline.

PHILIP GRAVES

Coastlines where the sea level in relation to the land level is stable tend to reduce in length. This is because headlands and the seaward side of islands erode. The sediment from this, and that carried by rivers running off the land masses, tends to fill in bays and estuaries, so islands become joined to the mainland as bars of sand or shingle (tombolos) form. This process is slower if the land is made of hard rock, faster if the land is sand or clay.

The almost straight coastline of East Anglia on England's east coast is a good example, with the Bure and Yare estuaries almost filled in, and coastal erosion planing off headlands. However, in the stable sea-level coast of southwest Scotland, the rock is harder. Erosion cannot smooth the coast as quickly.

Rising seas (or sinking land) create long, irregular coastlines of drowned river valleys (known as rias). South Cornwall and Devon in southwest England show this process well. Other drowned seashores such as the Greek islands or the Dalmatian coast also have long coastlines. Falling seas (or rising land) also create long, irregular coastlines because numerous undersea rocks and sandbars emerge. This is seen in the northern Baltic Sea.

Sea level changes (known as eustatic changes) are generally caused by global climate changes, as the sea itself expands in a hotter climate and ice sheets on land melt, increasing the volume of seawater. Changes in land level (isostatic changes) are often caused by crustal plate movements. The two effects sometimes combine. Climate change can cause crustal movement, such as when ice sheets melt and slide off the land, and the land consequently rises as the weight is removed. Volcanic eruptions can alter the land level, and they also alter climate by emitting CO_2 or ash, which warms or cools the planet respectively.

With the Earth's present water endowment, maximum global coastline length was probably reached a few thousand years ago at the end of the last ice age. River valleys in non-icebound regions would have cut down to a lower sea level, and were then ripe for the creation of long irregular ria coasts as the seas subsequently rose. Since then, many of these rias have filled in again, for example on the Sussex coast in southern England.

If you could remove enough water to lower sea level by 3,000 feet there would be no chance of a ria forming because no rivers have ever run that low on Earth and hence there would be no valleys to fill. In fact, there would be few rivers at all afterwards, because large parts of the continental interiors, far from the sea, would become desert. Global coastlines would probably be shorter than now.

Likewise if you added enough water to raise sea level by

3,000 feet the land masses would be so much smaller that even with plenty of rias, coastlines would be shorter. Moreover, the rias at this high sea level would be shorter because rivers this far above the current sea level would tend to have steep gradients.

HILLARY SHAW
School of Geography
University of Leeds, UK

One answer is the coastline has the same length whatever the depth of water over the Earth's surface. Imagine measuring the coastline using a long stick of unit length. You place one end of the stick at the water's edge and lay the stick down so that its other end is also at the water's edge. Assume the water is completely still. In between the two ends of the stick, the coastline wiggles either side of the straight line. So using this stick will underestimate the length of the coastline.

Now use a stick of half the length of the first stick. This stick measures the length of the coastline more accurately, but the coastline will still wiggle back and forth either side of it. This fractal behavior of coastlines leads to widely varying estimates of the length of a coastline. Indeed, zooming in to greater detail shows that the coastline continues to wiggle around either side of the measuring stick, however short it is, right down to the size of sand grains. This leads to estimates of the coastline length becoming longer and longer, with the most accurate measurement of the length of any coastline being almost infinite.

PETER WEBBER

Pingu's Pleasure

Why is it colder at the South Pole than at the North Pole?

T. P. LADD

Much of the temperature difference between the two poles can be explained by their difference in elevation. The North Pole (with monthly average temperatures in winter of around –22°F) lies on sea ice on the surface of the Arctic Ocean while the South Pole (at around –76°F) is 9,100 feet above sea level on the ice sheets of the Antarctic continent.

The background variation of temperature with height (in Antarctica about –18°F per mile gain in height) thus accounts for over half the difference. Also, the thinner (and hence colder, drier, and less cloudy) atmosphere overlying the South Pole reflects less heat back to the surface than its northern counterpart. Much of the remainder of the temperature difference can be explained by the contrasting atmospheric circulation regimes in the two hemispheres.

The continents of the northern hemisphere drive quasi-stationary "planetary waves" in the atmosphere. These waves transport heat polewards and also "steer" mid-latitude depressions into the north polar regions. The continents of the southern hemisphere are smaller and lower than those in the north, so the southern hemisphere planetary waves (and associated heat transport) are smaller.

The high mountains of Antarctica also block the poleward movement of mid-latitude depressions, which rarely penetrate into the interior of the continent. Finally, the atmosphere at the North Pole receives some heat from the underlying Arctic Ocean. Although the heat conducted through the 7 to 10 feet of sea ice that typically cover the ocean is small, large amounts of heat can be exchanged over the narrow "leads" of open water that occasionally form between ice floes.

JOHN KING
British Antarctic Survey, Cambridge, UK

Shrinking World

I once heard that if the Earth were shrunk to the size of a squash ball or racketball, the planet would be smoother than a real squash ball. Is this true? And if the converse happened and the ball were expanded to the size of the Earth, how high would the mountains be?

BY EMAIL

To answer this intriguing question, we first need to establish the scale factor we would have to shrink the Earth by in order to reduce it to the size of a squash ball. The Earth is 7,926 miles in diameter at the equator, and a regulation squash ball has a diameter of 1.7 inches. This means that to shrink the Earth down to the size of a squash ball, its size would have to be multiplied by a scale factor of 3.45×10^{-9}.

To compare the smoothness of the two surfaces, we need to know the variation of the surface—that is, the difference between the highest and lowest points.

For a squash ball, this is a simple process, because there are very few areas where the surface is higher than the average, but there are many small indentations or depressions. Since these depressions are roughly 0.004 inches in depth, the variation in surface height can be taken to be roughly 0.004 inches.

For the Earth, the lowest point below the surface is in the Mariana Trench, which is 36,200 feet below sea level at its deepest point, known as the Challenger Deep. The highest point is, of course, the summit of Mount Everest, which is estimated at 29,029 feet above sea level. Therefore, the variation in the height of the Earth's surface is 65,229 feet.

If we scaled the Earth down to the size of a squash ball, using the scale factor calculated above, the variation of its surface would be 2.25×10^{-4} feet, or 0.0027 inches. This figure is in fact about two-thirds of the figure for the squash ball, so what your correspondent heard is actually true: if

Earth were scaled down to this size it would indeed be smoother than the average regulation squash ball.

Now for the second part of the question. The lack of any raised areas on the squash ball's surface means that there are in fact mostly depressions or indentations in the surface. So if a squash ball were scaled up to the size of the Earth, there would be no mountains as such.

There would, however, be a lot of large craters. In fact, if we scaled up the indentations in the ball's surface we would end up with some immense depressions that were almost 18 miles deep.

If these depressions were ocean trenches like those that are found on the Earth's surface, they would penetrate the 3.7-mile-thick oceanic crust, and extend right through the Mohorovicic discontinuity where the crust meets the mantle and well into the mantle beneath. These craters would not only be deep but could be anything up to about 37 miles wide.

Tim Kelby

If the mass of the Earth were crushed down to the size of a squash or racketball, then it would be dense enough to be either neutronium or a black hole.

In the case of neutronium, the gravity at the surface would be in the order of at least a million times the gravity you are feeling now, more than enough to smooth out any irregularities in the surface. In the case of a black hole there wouldn't be a surface, just an event horizon which would be smooth.

Expanding a ball to the size of the Earth is somewhat different. If we assume that the ball is made mainly of carbon atoms and weighs 2 pounds, then there would only be around 352 atoms in each cubic inch. I believe that this is actually less dense than the Earth's upper atmosphere on the edge of outer space.

Stephen Forbes

Balance of Power

Is it true that Britain is sinking in the south and rising in the north? If so, why?

DAVE VALENTINE

Yes, it is true. It's the result of a process called isostatic rebound. Since the last ice age, a huge burden of ice has been removed from the north of Britain. Because the Earth's crust is not rigid, as it appears to us over a human lifetime, but very slightly elastic, it gradually responds to the addition or removal of weight above it by sinking or rising.

This adjustment takes thousands of years. If you remove, say, a layer of rock 1,000 feet thick from the crust, this will rise some 650 feet, just as removing cargo from a boat will make it rise higher in the water. Ice is about a third the density of crustal rock, so removing 1,000 feet of ice will cause the crust to rise around 200 feet or so.

Scandinavia and Scotland were under more than 1,000 feet of ice during the ice ages, and uplift here is fastest in the northern Baltic, where it still continues at nearly three feet per century. This is easily noticeable even over a human lifespan. The Hudson Bay area of Canada experiences a similar rate of uplift, for the same reason. In Britain the process is fastest in northeastern Scotland, where raised beaches exist several yards above the present sea level.

So why is the south of England sinking? First, the burden of Scottish ice pushed up the crust in surrounding areas that were ice-free, just as pressing down one part of a water bed makes adjacent areas of it rise. That process is now in reverse: the once-raised regions of southern England and the southern Baltic are now sinking.

Secondly, sea level is rising worldwide. Once it rose rapidly as the ice sheets over places like Scotland melted. Now global warming may be melting glaciers, sending more meltwater into the oceans. As the oceans grow warmer, thermal expansion also raises sea level.

So the south of England gets a double whammy—sinking crust and rising sea level. Without rising sea levels, the line between the sinking and rising parts of Britain would run between Wales and Yorkshire. Because of the latter, the line is further north, near the border of England and Scotland.

Regarding vulnerability to marine flooding, the London area is subjected to a quintuple whammy. Apart from the two factors above, the Thames Valley is a syncline, an area of locally subsided crust. Also, until recently groundwater extracted from below London was causing further subsidence. Finally, the funnel shape of the North Sea tends to bank up storm surges to ever greater heights as they enter the Thames Estuary.

All this adds up to one inescapable fact: the lower Thames was not a good place to site a major capital city.

HILLARY SHAW

As we sink, the sea is rising to engulf us. Also, thermal expansion of the oceans is increasing the depth of the sea by about 3 millimeters a year.

Here in Essex, in the southeastern corner of Britain, the net effect of sinking and rising water is a relative sea level rise of 6 millimeters a year. This is of huge concern to conservationists, coastal landowners, and those charged with maintaining sea defenses.

CHRIS GIBSON
Conservation officer
English Nature, Colchester, Essex, UK

Wave Goodbye

Over the best part of two bottles of wine my wife and I were arguing whether the stones we were throwing into the Mediterranean harbor of Ciutadella in Menorca were creating waves that would eventually hit the shores of North America. She argued they wouldn't, having to pass through the Strait of Gibraltar and cross the Atlantic, while being assaulted by other waves, friction from the coast and seabed, and storms. I said that until they encounter a shore, waves are virtually infinite. Who is right?

DAVE JOHNSTONE
Hounslow, Middlesex, UK

Your wife is right. Waves traveling in a fluid are not infinite. They lose energy because the passing of a wave causes an upwards and downwards displacement of the water, and hence there is always an associated dissipation of energy due to viscous forces.

In addition, as the wave spreads, its energy gets distributed out over a greater and greater perimeter, causing the energy density to drop until it would become unmeasurable against background noise. Waves traveling in deep water do not cause a very large displacement, and so the rate of energy dissipation is relatively slow. This is why tsunamis can travel such great distances. However, in a shallow puddle or even the relatively shallow waters of the Mediterranean, energy is dissipated much more quickly.

SIMON IVESON
Pembangunan National Veteran University, Indonesia

Ripples spread out in an ever-increasing concentric circle. Making the rather unrealistic assumption that the height of a ripple is directly proportional to the energy that creates and sustains it and that there are no losses to friction, the height of these ripples is inversely proportional to their radius. This is because the energy associated with the ripple at any point has to be shared around the whole circumference of this ever-enlarging circle.

So, by the time the ripple passes through the Strait of Gibraltar 620 miles away, its radius has increased a million-fold and its amplitude would be a million times smaller than when the circular ripple had a radius of only 3 feet. A ripple 4 inches high would already be an imperceptible 100 nanometers high as it left the Mediterranean.

In principle, ripples should be able to pass through other waves undisturbed. But, if the questioner's ripple were to break onto an American beach 4,000 miles further away, it would only be about 10 nanometers high, equivalent to the thickness of 100 atoms. Even this height would not actually be achieved because of air friction and the viscosity of the water itself. To make matters worse, the Strait of Gibraltar is not on a line of sight from Ciutadella so the ripple would need to be reflected off North Africa or a passing ship and this would involve an even greater loss of energy.

Waves normally have a source of energy—the wind—to sustain them. The ripples from a stone would have to survive on the tiny morsel of energy as you tossed them. To make matters worse, the rate at which a wave loses its energy is inversely related to its wavelength so, with their tiny wavelengths, your ripples quickly lose the small amount of energy they start with. On the other hand, a tsunami such as the disastrous one that occurred in the Indian Ocean in late 2004 is created by the release of a huge amount of energy associated with events like submarine earthquakes. They have typical wavelengths of 300 miles, so they lose very little energy as they head for distant landfall. In deep water they can outrun a jet aircraft, but with a wave height of a yard or so, they go unnoticed aboard ships. However, when they reach shallow water they slow down and start "shoaling," often reaching

many feet in height and sometimes traveling far inland. This is probably the scale equivalent of dropping stones into a rock pool.

Mike Follows

Have you ever seen the waves in a still pond die out a short while after you've tossed in a pebble? The viscosity of the water will dampen the wave you started so that it dies out long before it reaches the Straits, unless you toss in a truly gigantic stone.

Morton Nadler

Even if an initially circular wave front encounters no shores, the height of its wave crest diminishes in inverse proportion to the square root of its distance from its origin.

Initially, the entire wave formed by a small rock dropped directly down into deep water has a circular pattern. But this pattern is broken up if segments of the wave front are reflected from various shorelines at various angles and times.

If a shoreline is very shallow and sandy, or marsh-like, practically all the energy of such a wave's front segment reaching it is apt to be absorbed by the elements of the shore, and so there is no significant reflection. Near-perfect reflection in a single main direction without significant energy loss can only be expected when a segment of the wave front strikes something like a smooth, hard cliff side in deep water. But if the horizontal contour of the cliff side at water level is jagged rather than smooth at the scale size of the wave crest, the wave segment becomes scattered, with fragments being reflected off in different directions.

Looking at detailed maps of the Mediterranean, it is obvious that no segment of a wave front originating from a rock dropped in the harbor of Ciutadella in Menorca could travel to the Strait of Gibraltar and out into the Atlantic Ocean without first striking a good number of shores and being reflected by each.

Another factor is wind. Most surface waves in bodies of

water are created by the wind. If a wave front created by a rock falling into water travels in the same direction as a gentle wind, it may become larger. But a strong wind can completely deform it to the point where its identity becomes lost.

It seems unlikely that any significant part of the original wave from a rock dropped in Ciutadella harbor would ever reach the Strait of Gibraltar, let alone North America. But I appreciate the spirit of the question.

PATRICK JOHNSON

6 Weird Weather

No-Ball Snow

On vacation in Scotland last winter, I found it impossible to make snowballs with that day's newly fallen snow. It was an extremely cold day and it would not stick together in the way it normally does. My friend has also experienced this in the Alps. What is the reason for it?

MORAG CHALLENOR

It's not surprising that this question comes from someone in the United Kingdom. No one from Canada or the northern United States would write that snow "normally" sticks together in a way suitable for making snowballs. All Canadian and many North American children know that sometimes snow is good for "packing" and sometimes it isn't.

From my memories of a North American childhood, the relevant variable is temperature. When the air temperature is only a little below freezing, as it often is when snow falls in the UK, the snow is usually wet, comes in big flakes and is good for packing. When it is really cold, say about –4°F, the snow is usually dry and powdery and is no good for packing. Presumably the moisture content of the snow determines the amount of ice that forms under the pressure of the snowballer's hands, and it is this ice that makes the snow stick together.

A colleague who also grew up in the cold parts of North America reminds me that when the temperature is too far above freezing, snowballs simply turn to slush as you make them. So there is an optimum band of snow temperature for packing, and it just happens that snow in the United Kingdom generally falls within that band.

Bob Ladd

Only wet snow, containing up to 50 percent liquid water, is good for making snowballs, and this needs temperatures around freezing point.

In 1842 Michael Faraday suggested that wet snow has a thin film of water on the ice particles, and that this is responsible for gluing them together. He suspended two blocks of ice in a bath of freezing water to show that simply bringing them into contact was sufficient to make them stick together.

Lord Kelvin had a different explanation. Squeezing the snowball brings the points of the ice crystals into contact. Although our hands cannot exert much pressure, the local pressure at the sharp points of the ice crystals can be high enough to cause melting. The instant this pressure is released, the water freezes again. However, the colder the snow, the higher the pressure required.

Our understanding of surfaces is now more advanced. Water molecules on the surface of ice particles are not bound to anything on the air side, so they have excess energy. This energy can be reduced if two surfaces come together, just as Faraday observed.

But if this were the whole story, we would also be able to make snowballs at temperatures well below freezing. At very low temperatures, snowflakes, which come in all shapes and sizes, do not fit together snugly. However, at temperatures closer to freezing, individual water molecules become more mobile and migrate over the surface to fill the awkward gaps between flakes. This ensures that neighboring flakes fit together much better. With a bigger contact area between flakes, they now stick together more readily.

Mike Follows

Which Way, Captain?

In the novel Moby-Dick, *the wooden whaling ship meets a typhoon southeast of Japan and is subjected to thunder, lightning, and displays of St. Elmo's fire. Subsequently, the magnetism of the ship's compass needle is discovered to be reversed. Author Herman Melville maintained that such compass reversals "have in more than one case occurred to ships in violent storms," and sometimes when the rigging has been struck by lightning the magnetism in a compass needle may be totally lost. Is this fact or fiction, and if true how does it occur?*

ALAN SLOAN

Herman Melville's assertion is entirely plausible. Lightning involves very high currents with high associated magnetic fields. They remagnetize exposed outcrops of high-coercivity (high resistance to the effect of an applied magnetic field) rocks with ease. Currents exceeding 10,000 amperes have been deduced from rock magnetizations. The associated magnetic fields could easily demagnetize or reversely magnetize a compass needle.

ALAN REID

A moving electrical field will induce a magnetic field, and an electrical discharge such as lightning can easily cause a compass needle to lose or reverse its magnetization.

What the questioner failed to mention is that Captain Ahab fashioned a new compass by striking a sailmaker's needle and thereby magnetizing it. This is rooted in fact. I have demonstrated this phenomenon more than once by accidentally dropping a pair of expensive tweezers used for handling metallurgical samples. The fall was sufficient to magnetize the tweezers.

Ferromagnetic materials are composed of microscopic magnetic domains, which may be oriented in random directions producing a demagnetized state. By aligning the domains in more or less the same direction, the material becomes magnetized. In some cases, a sharp blow will impart enough energy to make this happen.

Roger Ristau
Institute of Materials Science
University of Connecticut, Storrs

Ice Art

On some cold mornings the frost on windows and cars makes patterns that look just like leaves, ferns, and branches. How does this happen?

Bob Clarke

Waking up to frosty bedroom windows is becoming a thing of the past, thanks to the insulating properties of double glazing and cozy central heating. But if you are still stuck with single glazing, on winter mornings your view will be obscured by fern-like patterns of frost.

Panes of glass lose heat quickly on cold nights, cooling the water vapor molecules in the indoor air nearest the glass. The temperature of the water molecules in the air can fall below 32°F without them actually freezing. But as soon as this supercooled water vapor touches the cold glass, it turns directly to ice without first becoming water.

Tiny scratches on the surface of the glass can collect enough molecules to form a seeding crystal from which intricate patterns then grow. Up close, the crystal surface is rough with lots of dangling chemical bonds. Water vapor molecules latch on to these rough surfaces and crystals can grow quickly. The structure of the elaborate branching depends on both the temperature and humidity of the air, as well as how smooth and clean the glass is. When the air is dry, the water molecules condense slowly out of the air and cluster together in stable hexagons. The six straight sides of these crystals are relatively smooth with very few dangling bonds, giving water vapor molecules little to hang on to.

Feather-like patterns are more likely to form on clean windows and when the air is heavy with water molecules. Under these conditions, lots of water vapor molecules bombard the seed crystal and there is no time for the stable hexagons to form. Instead, the molecules latch on to the dangling bonds that stick out of any bumps in the crystal, which means the bumps grow even faster. These bumps eventually grow into large branches, and in turn the bumps on the branches become lacy fronds.—Ed.

Heavy Weather

Is it possible for a ground-based observer to calculate how much water is in a particular cloud? Do size and color have any bearing? If this is possible, how is it done—I want to impress my friends. If it isn't, can it be calculated by more scientific methods?

James Down

As explained below, there is an approximate method to answer this question that will impress your friends, and a precise answer for which you'll need an excellent Doppler radar system and a large grant from the government.—Ed.

The amount of water inside a cloud is not different from the amount of water in the clear air around it. However, while the clear air contains water vapor, inside the cloud the air is saturated with water vapor and it has condensed out to produce the cloud. The difference between the two states is caused by temperature differences rather than the water content.

The color of a cloud doesn't make much difference either. In the higher part of a cloud, the water is in the form of ice crystals. Lower down it is a mixture of ice and liquid water. The color of the cloud depends mainly on this ice/water mixture and the size of the water droplets, and less on the total amount of water.

An estimate of a cloud's water content can be obtained from the amount of rain the cloud can produce (see Marcel Minnaert's *The Nature of Light and Color in the Open Air)*.

If the entire atmosphere were saturated with water and it all fell in a steady stream, this could produce some 1.4 inches of rain, while the thickest actual clouds produce about 0.8 inches.

Cloudbursts can produce 2 inches or more, but this requires additional moisture from the surrounding atmosphere, which means such events are localized.

The heaviest cloudbursts roughly follow the following equation: the rainfall in inches equals 162.5 multiplied by the square root of the time in minutes that it has been raining. A more typical shower produces a fraction of an inch of rain, at a rate of perhaps 0.04 inches per minute. Normally 1 inch of rain corresponds to 900,000 cubic feet of water weighing 4,000 tons per cubic miles of cloud, though the thickest clouds can contain up to 20 times as much.

You can also estimate the amount of water from the volume of the cloud. By volume, the fraction of the cloud filled with water is about 1 millionth, or 0.0001 percent. The cross-section area of a cloud can be measured from its shadow. A small cloud 1,000 feet by 1,000 feet and 500 feet high has a volume of 500 million cubic feet, of which roughly

500 cubic feet will be water, weighing 15 tons. Even if you can't calculate the precise amount of water in a cloud, these numbers may still impress your friends.

ALBERT ZIJLSTRA
Department of Physics
UMIST, Manchester, UK

Sadly, just looking at a cloud does not give very precise information about how much water it contains. The color of a cloud depends entirely on the relative position of the viewer and the physical structure of the cloud. Its apparent size is dependent on the altitude of the cloud, and this is generally very difficult to judge from a single observation point.

But knowing exactly how much water is contained in a cloud is important for producing accurate weather forecasts.

The choice of frequency for a Doppler radar beam is very important. If the beam interacts too strongly with the water in a cloud, in terms of either reflecting or attenuating the signal, then the radar will have only a limited ability to penetrate cloud structure. If the interaction with water is too weak, then no useful information can be returned at all.

One facility in England can analyze and extract a huge variety of data and has a maximum range of around 100 miles. It is able to provide information on droplet density, size, speed, and whether the droplets are water or ice.

Using this tool, your reader could work out fairly accurately the total water content contained within a cloud and, from the structure of the cloud, how likely it is to start raining—this technique has proved very useful at the Wimbledon tennis championships in past years, which are, of course, notorious for being interrupted by downpours.

Such radars help to produce detailed information on weather, from tracking hurricanes to helping to produce your daily weather forecast and predicting areas of turbulence on aircraft flights.

DAVE RICHARDS

Heavy or Light

What causes different types of rain? Sometimes it comes down in "stair rods"—lengthened droplets that fall at great speed and bounce high after hitting the ground. Other times there's just a misty drizzle that blows aimlessly in the breeze. How can rain fall so heavily that it can cause physical pain, or so lightly that it is just a soaking mist? And how do you get the types in between?

MARTIN REEVES

Elongated stair rods are an illusion. Large drops actually tend to be flattened by air resistance. When they land they are called in Afrikaans (and I understand, in Welsh) "old women with clubs." The circular sheet of splashing water suggests a wide skirt and the centrally rebounding droplet a cudgel.

Droplet size is the main factor in creating different kinds of rain. This depends on conditions at the time of formation, particularly humidity, temperature, and the number of air-borne nuclei such as dust particles. For example, moderate numbers of nuclei in moist updrafts tend to promote drops that grow large, because there is plenty of water for them and they cannot come down before they grow heavy enough to fall faster than they are lifted. When nuclei are crowded they compete with each other and can only form small droplets that may evaporate before they reach the ground.

In still air, big drops fall fast and hard. Drops that are about 0.4 inch in diameter reach speeds of around 20 miles per hour, at which point their own slipstream tears them into smaller droplets, unless they are partly frozen. This limits raindrop size. However, a large number of falling raindrops can create a downdraft, increasing the downward velocity a drop can achieve without splitting, while strong horizontal winds can more than double the speed of impact.

And remember that kinetic energy rises with the square of the velocity.

JON RICHFIELD

Rainfall intensity depends mainly on the depth of the cloud and the strength of the updrafts. Rapidly rising air produces fast condensation of water droplets and large amounts of rain, mostly when the cloud extends high enough for ice crystals to form among supercooled water droplets.

Shallow clouds with weak updrafts only give drizzle, which rarely falls faster than 10 feet per second. Large raindrops can reach a terminal velocity of about 32 feet per second. Their fall speed increases with size until the diameter approaches 0.25 inches, at which point wind resistance flattens the base, increases the drag and prevents further acceleration.

However, if the rain is caught in a "downburst" where an air column is descending at 65 feet per second or more, the rain hits the ground harder. Downbursts are often associated with cumulonimbus clouds that contain almost vertical air currents. The weight of precipitation in the cloud may be enough to trigger a downburst.

Rain from deep flat cloud layers is usually caused by slow diagonal ascent along a sloping frontal surface. Such rain is persistent but seldom heavy. This can change if prolonged lifting makes the layer unstable. Then massive turrets containing strong updrafts grow vertically out of the layer. These can produce heavy downpours from a cloud mass which had previously only given moderate rain.

TOM BRADBURY

Forest of Fear

I read recently that oak and fir trees are far more likely to be struck by lightning than pines, and beech is the tree least likely to attract a lightning strike. Does anybody know if this is correct and, if so, why?

JEFF KESSLER

Almost a century ago, a monthly journal, *Country Queries and Notes*, was first published. This question was raised in its first issue and attracted an enormous amount of interest from readers. Their responses were still arriving a year later.

Replies were compiled from many different sources and showed the following number of lightning strikes: oak, 484; poplar, 284; willow, 87; elm, 66; pine, 54; yew, 50; beech, 39; ash, 33; pear, 30; walnut, 22; lime, 16; cherry, 12; chestnut, 11; larch, 11; maple, 11; birch, 9; apple, 7; alder, 6; mountain ash, 2; and hawthorn, 1. Only after further prompting from the editor did one reader admit to seeing a stricken sycamore. There were assertions that holly was never seen to be hit.

With no indication of the relative proportions of the trees present the list may appear meaningless, but it does look as if height has more relevance than species. Arguments grew a little wearisome but there was some support for the view that trees with corrugated bark (and therefore traces of moisture) attracted more strikes than those with smooth bark.

It seems fairly clear, though, that it is best to assume what we all know already—lightning can strike anywhere.

JAMES O'HAGAN

The Forestry Commission's most recent research paper on the subject states that oak, poplar, and Scots pine are the tree species most frequently damaged, with beech least likely to be damaged. However, this is based on two surveys, each with relatively few records. The first was carried out between 1932 and 1935 and the second between 1967 and 1985, and there are some discrepancies, partly because the earlier survey included only obviously lightning-damaged trees, while the later one includes noncatastrophic damage, which is quite common and sometimes barely detectable.

North American publications also show beech, birch, and horse chestnut as relatively unlikely to be struck, compared with oaks, pines, and spruces, among others. North America gets much more lightning than Britain and it is not uncommon for particularly valuable specimens to be fitted with lightning conductors in order to preserve them. To my knowledge a conductor has only once been fitted to a British tree, a particularly large cedar that was considered especially at risk.

Various theories have been advanced as to why some species are apparently struck more often, including the possibility that some conduct electricity better, either through the sap or because their rough bark retains more rainwater. This would explain why beech, with its smooth bark, is less susceptible to strikes.

SIMON PRYCE

Knowing Your Dews

There have been many times that I have unzipped the flap door on my tent while still in my sleeping bag to see that a heavy dew had fallen during the night. Being so close to the dew-laden grass, I always notice that the individual drops occupy an apparently precarious position at the very tips of grass blades. How do they get there and how do they stay there?

JOHN LAMONT-BLACK

This process is called guttation. On the surface of leaves there are stomata or pores through which water is lost by transpiration. At night, the stomata close, causing a reduction in transpiration. Drops of water are then forced out of the leaf through special stomata or hydathodes. These special stomata are found along the edges of the leaves or at the tips. It is believed that guttation is caused by high root pressure. Grasses often force water out of the tips of their blades, as your wide-awake camping correspondent noticed. Guttation also happens in potatoes, tomatoes, and strawberries on their leaf margins.

FRANCES TOBIN

The drips result from guttation. Plant roots take up inorganic ions from the soil and transfer them to the xylem from which they can't leak back. Water is drawn in by osmosis, which creates positive pressure in the xylem. Because of this pressure, xylem sap leaks from pores (the hydathodes) at the tips of grass leaves (or directly from the cut or cropped ends of the leaves). When the drops are large enough they fall and new ones form.

Guttation usually happens at night because, by day, water loss from the leaves is normally sufficient to maintain a negative pressure in the xylem. Conditions favoring guttation also favor camping: clear skies and light winds, warming the ground by day and cooling the air (thus raising the relative humidity) by night, with the ground moist enough to accept a tent peg.

Some useful ions are probably recovered by the hydathodes and some of the ions in the xylem in the roots may have been recirculated by the phloem. A similar process to guttation brings calcium to developing fruits. Its interruption is bad news. For example, a dry atmosphere in a greenhouse by night can prevent positive pressure building up in the xylem, and developing fruits such as tomatoes can show calcium deficiency in the form of blossom-end rot.

JOHN TULETT

Guttation has been observed in more than 330 genera of 115 plant families and is induced by conditions which permit root absorption but retard transpiration. It is, therefore, more frequent at night and commonest in humid tropical areas where the higher soil temperature favors root absorption and the moist atmosphere retards transpiration. Among plants in temperate regions, species of Impatiens and grasses, including cereals, readily exhibit guttation. In the tropical plant *Colocasia antiquorum* 7 ounces of water may be exuded by a single leaf in just one day.

JOHN TOMLINSON

7 Troublesome Transport

Wrap Up Well

Recently I was flying at 39,000 feet and 500 miles per hour through air at a temperature of 122°F. The wind chill should have been horrendous, but luckily I was in an airliner. However, the cabin walls were only 4 inches thick. What insulation is used, and can I have some for my sea-level home?

PHILIP WELSBY

The wind-chill factors one commonly sees are for turbulent flow, usually over exposed human skin, which loses heat by evaporation and convection. Laminar flow over smooth dry metal, which is what aircraft designers aim for, is much less efficient at transferring heat. At 29,500 feet, air density is about one-third of its value at sea level: it is as if the plane were flying in a vacuum flask.

At speeds above 300 miles per hour, there is significant frictional heating of the outer surface of any aircraft. Parts of Concorde got 390°F warmer in flight, and the skin of a returning spacecraft of course gets red hot.

The human metabolic power density in watts per cubic yard inside an aircraft packed with passengers is hundreds of times higher than in even the smallest of houses, and the surface-to-volume ratio of a smooth cylinder is much less than that of an irregularly shaped house.

The air inside a pressurized cabin is temperature-conditioned and circulated. Several megawatts of surplus heat is available from the engines to drive the air conditioning system, so there's no problem maintaining a comfortable cabin air temperature in flight. The trick is to line the cabin with just enough plastic so you can't touch any cold bits of metal, and to fill the cavity between the skins with fairly ordinary foam or fiber insulation, with properties very similar to the walls of a house. So it's pleasant in the cabin, but your correspondent is right when he assumes that parts of the aircraft do chill. The tail cone and rear baggage compartment do indeed get very cold on a long flight.

It's worth noting that a plane on the ground with the engines switched off is no warmer than an unheated trailer.

ALAN CALVERD

The temperature outside a plane at high altitude is very low but the aircraft's skin can become very hot. These extreme temperatures require thermal insulation but, apart from a couple of areas such as those above and below the passenger compartment, the main insulation consideration is acoustic.

To keep the inside of the plane protected from engine and wind noise, the insulation is thicker than that required for thermal considerations. Fiberglass batting, the material used, has the very small fiber diameter necessary to provide the best acoustic insulation. The insulation usually varies between 5 inches in the roof to 3 inches in the sides and 1 inch in the floor, with obvious variations between different types of plane. The type of fiberglass used in planes is especially light, but similar products are widely available to the public and commonly used in buildings for thermal insulation.

DAVID KETTLE

Lighting Up

When you go to the toilet on an airliner, bolting the door operates the light inside the cubicle. However, after you complete the circuit by sliding the bolt, the light takes a couple of seconds to come on. Why does it do this?

MICK TOWNE

It takes a couple of seconds for the low-power, low-voltage fluorescent light fixture to supply enough energy to the tube to establish an arc across (or through) the gas used in the light fixture. This is because the small quantity of mercury that is used to create the arc needs to heat up just enough to start ionizing.

The mercury arc, which produces light at a wavelength of 337.1 nanometers just outside the visible spectrum in the ultraviolet range, excites the atoms in the coating inside the tube and they then emit photons in the visible range. The delay happens as the arc is being established.

Unlike the electricity supply that powers modern fluorescent light fixtures in the home and office, the 24-volt electrical systems in aircraft are less "stiff," meaning they generate less energy. But the surge needed to light an instant-start fluorescent tube would create a significant voltage dip, lasting a few tens of milliseconds, that could play havoc with other aircraft devices. This is the reason the light has a "soft start."

DAN SCHWARTZ

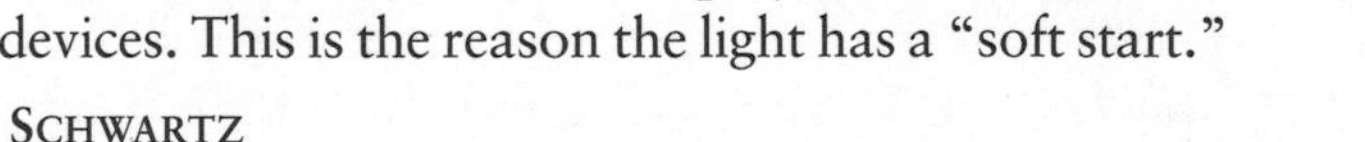

Pre-Inventing the Wheel

Wheels are a pretty effective method of getting around. Is there any reason why they never evolved in nature?

TYRONE PEELER

It is not true to say that nature hasn't invented the wheel: bacteria have been using it to get around for millions of years. It is the basis of the bacterial flagellum, which looks a bit like a corkscrew and which rotates continuously to drive the organism along. About half of all known bacteria have at least one flagellum.

Each is attached to a "wheel" embedded in the cell membrane that rotates hundreds of times per second, driven by a tiny electric motor. Electricity is generated by rapidly changing charges in a ring of proteins that is attached to the surrounding membrane. Positively charged hydrogen ions are pumped out from the cell surface using chemical energy. These then flow back in, completing the circuit and providing the power for the wheel to rotate.

The only nutrients that the flagellum needs are protein building blocks to allow it to grow longer. These are forced up through the hollow center of the flagellum and are assembled into new flagellar material when they reach the end.

It is a very sophisticated piece of nanotechnology and even has a reverse gear that helps the organism find food. So, far from nature not having invented the wheel, given the very large number of bacteria in existence, there are probably more wheels in the world than any other form of locomotion.

ANDREW GOLDSWORTHY

There is one form of macroscopic life that rolls like a wheel: tumbleweed. The above-ground part of the plant detaches from the stem and rolls in the wind, dispersing its seeds.

ERIC KVAALEN

So nature has created a wheel and exploits rotation. Why don't creatures make more use of them?—Ed.

Wheels are only effective because we have modified so much of our habitat to accommodate them (though not quite enough, as wheelchair-users will tell us). Once you move off the roads and paths and try to traverse the rough, wet, muddy, boggy, rocky, sandy, snowy, icy, steep, or fissured terrains that cover most of the world's land masses, our brilliant invention becomes little better than useless.

ALISTAIR SCOTT

Every single part of our body (and those of most other developed organisms) is attached to, or communicates with, the systems that regulate our body—the main ones being the nervous system and the blood system. The same would need to be true of any wheel developed by an organism. But wheels have to be free to rotate. If they were attached to an organism, they would wrap blood and nerves around the axle.

BEN HILL

Wheeled machines such as cars and trains rely on an engine that generates torque. In the case of wheelbarrows and buggies, leg power is required. Most animal locomotion is produced by muscles which are good at converting chemical energy into work, but can only contract. To use wheels, nature would need to do one of three things: replace the muscle with a different source of thrust which would be a complex evolutionary process; combine wheels with legs which rather defeats the purpose; or arrive at something improbable such as a bionic bicycle.

ROLAND DAVIS

Evolution is not a process that "thinks ahead." It is merely the cumulative effect of natural selection acting on the results of chance mutations. Hence any new forms of life or locomotion only emerge if every single intermediate step conveys some sort of competitive advantage to the organism, or at the very least does not convey any disadvantage.

Therefore wings can evolve because a partial wing conveys some aerodynamic benefit to an animal leaping between tree branches. Similarly a hard shell can evolve because a

partial shell conveys at least some protection. However, it is difficult to think of an intermediate stage to a wheel that would offer any competitive advantage instead of being an unwanted burden.

Of course, in reality wheels have evolved: evolution resulted in the emergence of human beings, who have been clever enough to make wheels, and to subsequently alter the environment around them to be suitable for their use.

SIMON IVESON
Universitas Pembangunan Negeri, Indonesia

Sea Legs

When I returned home after a day of sailing lessons, I still had the feeling that the room was moving up and down. Why is this?

RICHARD MATTHEWS (AGE 9)

In order for you to estimate your location, your brain combines information from a variety of sources, including sight, touch, joint position, the inner ear, and its internal expectations. Under most circumstances, the senses and internal expectations all agree. When they disagree, there is imprecision and ambiguity about motion estimation, which can result in loss of balance and motion sickness.

On boats, seasickness may develop because of conflict between sensory input and internal expectations about motion. Developing "sea legs" is nature's cure for seasickness: you become accustomed to anticipating the boat's movements and prepare to adjust your posture accordingly. When you finally go ashore, you may feel your body continuing to do this for hours or even days afterwards, making it seem as if the room is moving and sometimes even leading to nausea.

A few unfortunate people experience persistent symptoms lasting months or even years. This is known as mal de debarquement syndrome. Exactly why their symptoms persist so long isn't understood, but they can be treated.

Sailing isn't the only activity that causes illusory motion after-effects. Overnight rail passengers sometimes say they can feel the "clickety-clack" of the track in their legs after they leave the train. And astronauts returning to Earth commonly experience vertigo, nausea, difficulty walking, and sensory flashbacks. The longer one is exposed to the unfamiliar motion, the more prominent and long-lasting are the after-effects.

TIMOTHY HAIN
Department of Physical Therapy and Human Movement Sciences, Northwestern University, Chicago, Illinois
and
CHARLES OMAN
Man Vehicle Laboratory, MIT, Cambridge, Massachusetts

Titanic Explosions

A few years ago I attended the Titanic exhibition at London's Science Museum. One of the exhibits informed me that great care had to be taken when bringing cast-iron objects to the surface from 3 miles down on the seabed, because when they emerge from the water they can explode. Why do these objects do this?

THOMAS THEAKSTON

There are several phenomena involved. One is that cast iron invariably contains small gas cavities or blowholes that are formed well beneath its surface. Another is that it has quite low ductility, and will fracture rather than deform. Thirdly, it is a very heterogeneous material, containing about 4.5 percent carbon and significant amounts of silicon and manganese, together with phosphorus and sulphur. The principal phases that are present are graphite, argentite, and ferrite.

When immersed in an electrolyte such as seawater, electrolytic corrosion starts up at the surface of the casting. One of the products of this corrosion is hydrogen in an ionic or atomic state. In this state it can diffuse through the ferrite lattice and find its way to the gas cavities. There it re-forms as molecular hydrogen, increasing the pressure in the cavities.

Because this electrolytic process takes place at great depth and pressure, the pressure build-up in the gas cavities reaches equilibrium with the external water pressure. Raising the cast-iron object from the deep seabed removes the external pressure on the iron, so the gas in the cavities creates very high stresses.

At best, the iron will develop cracks. At worst, the casting will shatter.

C. C. Hanson
Metallurgist

Old cannon balls brought up from the sea sometimes explode after being handled. This happens under special circumstances, when sulphate-reducing bacteria that are common in ocean sediments colonize the minute cracks and crevices in the iron. The bacteria use sulphates in the seawater as a source of oxygen and excrete the resulting reduced sulphur species. In the presence of iron, the soluble sulphur species react to form iron disulphide (pyrite) or iron monosulphide minerals.

Iron sulphides, thermodynamically stable under the reducing conditions on the seafloor, commence oxidation as soon as they are brought to the surface. This reaction is highly exothermic, produces acid, and involves a considerable increase in volume. Substantial oxidation can occur within hours, perhaps even faster. Within confined spaces, the rapid volume change of brittle objects during oxidation can result in potentially explosive break-up.

JEFF TAYLOR
Principal environmental geochemist
Earth Systems, Kew, Victoria, Australia

Hail the Ale

When Britain ruled India, brewers developed a special type of beer known as India pale ale (IPA) that, after being brewed in Britain, could withstand crossing the equator twice before arriving in India still in a drinkable condition. This was in the days before beer could be pasteurized and filtered, and so it was heavily hopped to keep it in good shape. What are the properties of hops that help them to preserve beer?

B. MANDERS

Hops contain a group of compounds called humulones or alpha-acids, that are very insoluble in water, but which undergo a chemical rearrangement during the wort boiling to form an isomeric group of compounds called isohumulones that are water soluble.

The analogs of humulone (and isohumulone) also found in hops—cohumulone, adhumulone, and prehumulone—differ only in the number of carbon atoms in the side chains. It is these iso alpha-acids that give beer its characteristic bitterness and also exert "bacteriostatic" effects on most gram-positive bacteria. In other words, they do not kill bacteria, but prevent their growth.

Before the advent of refrigeration and pasteurization, the only way to prevent spoilage in beer was by using alcohol and hops. Alcohol provided an unfavorable environment for

microbial action, and the isohumulone content of the hops inhibited the growth of bacteria such as *Lactobacillus*. Thus, high alcohol content (in German export beers, for example) and high hopping rate (as in India pale ale) could protect beer from the souring associated with long storage times.

India pale ale was invented by George Hodgson, a brewer at the Bow Brewery in East London in the 1790s, who took his pale ale recipe, increased the hop content considerably, and raised the alcohol content by adding extra grain and sugar. Hodgson also added dry hops to the casks at priming, when sugar is added to allow secondary fermentation, and conditioned the beer with more sugar than was typical for pale ales. The high sugar priming rate probably helped keep the yeast alive during the voyage and resulted in a very bitter, alcoholic, and sparkling pale ale that could withstand the rigors of travel while having a reasonable shelf life in India.

Stefan Winkler
Vice-Consul Science and Technology
British Consulate-General, Boston

With such large amounts of hops and alcohol, what did it taste like? Hops are the most expensive ingredient in any beer and a modern brewery probably couldn't brew a genuine nineteenth-century India pale ale economically. With so many hops, a young IPA would hit the cheek cells like paint stripper. However, the long voyage and the pale ale futures market meant the beer usually spent 12 months aging. This turned the hop bittering from an aggressive taste to a fine bouquet, which some writers described as reminding them of a French white wine.

I found this description fanciful, but I have carried out research on IPA. In the brews I made, especially those based on Edinburgh recipes of around 1840, the taste after a year was unlike anything currently on the market. It certainly wasn't overpoweringly bitter.

The combination of hops and alcohol provided a very

powerful antibacterial environment, but there was still much that could go wrong. The ale was vented before the long voyage to prevent serious explosions during the crossing and this may have introduced bacteria. It was also not easy to sterilize the casks before filling them. However, it was a trade worth pursuing because of the huge volume of empty cargo ships returning to the colonies and the cost of carriage was very low.

CLIVE LA PENSEE
Author of *Homebrew Classics: IPA*

Many brewers who exported porter (or Imperial stout) to Russia during the nineteenth century increased the beer's life by boosting its hop and alcohol content. These dark porters, brewed in England but popular with Russian royalty, were high in alcohol, sweet, and dense, and survived the journey from Britain to the Baltic and across Russia. They are still popular in the Baltic states, where they are still brewed. Samuel Smith's of Yorkshire produces a fine example.

At the same time as IPAs were becoming popular, the first golden lagers of Plzen (known as pilsners) were spreading fast, making use of the new railways and liberal doses of Saaz hops.

Most IPAs tend to be bitters that are hoppier than the norm, although a handful of "historical IPAs" weighing in at 6 percent or more alcohol by volume (still less than the 10 percent plus of the originals) are brewed by such microbreweries as Burton Bridge and Freeminer.

LAURENT MOUSSON

Tread Mills

Why do car and motorcycle tire manufacturers keep coming up with such varied tread patterns? Every time I look

at a tire, the design seems different. Why is there no standard, proven pattern?

G. CURLING

There are only two physical requirements for a car tire tread pattern. It must provide traction for acceleration and braking and it must move surface water out of the way so that the tire can touch a wet road without aquaplaning, which is where a vehicle slides uncontrollably on a wet surface.

A simple block pattern is great for off-road traction, but the front and back of each block wears quickly on tarmac. Circumferential ribs, edged with tooth-like indentations, give extra traction without increasing the overall wear. Regularly spaced cross-tire features, however, generate a loud noise so irregular patterns are used.

At 60 miles per hour in moderate rain, a car tire has to displace five liters of water every second just to maintain contact with the road. Crosswise cuts in the tire lift water off the road, which is then squirted out sideways through tunnels in the side ribs.

With motorcycle tires, the oval cross-section of the tread cuts naturally through water, so aquaplaning is rarely a problem, and noise levels for the rider are hardly an issue given all the other noise. All that is needed is traction.

It is clear that these requirements can be fulfilled by many different patterns. In fact, most variations in tread design are decided by the tire makers' marketing people.

REINHARD READING

In the late 1980s I worked on three-dimensional CAD/CAM design software for a top tire manufacturer.

The software allowed the designers to create almost life-like pictures of tires based on two-dimensional drawings of the tire section and the treads.

The designers told me that this saved them a lot of time, because many of the hundreds of designs they produced every year were turned down by the marketing department purely because of their appearance. Tread patterns were described as "not sexy enough" or "not masculine enough" and were sent back to the drawing board.

Only when the marketing department agreed with the general look of the tire and its tread pattern could a set of test tires be produced. These were cut by hand, because of the high cost of a mold, and then tested.

Experienced tire designers knew what made a certain tread pattern perform well for a given type of tire and could therefore deliver surprisingly new and different designs again and again that satisfied the marketing department's desire for new products but also still behaved well during the tests.

ANDRE DE BRUIN

Ship Shifting

Suppose a large ship, such as the QE2, *is floating freely alongside a quay and no forces such as wind or sea currents are acting on it. If I stand on the quay and push the side of the ship, will it move, even very slowly and slightly? Or is there some sort of limiting friction caused by all those water molecules around the hull that can only be overcome by a much larger threshold force?*

TREVOR KITSON

While I was a conscript in the service of King George V, on several occasions I moved a destroyer under the circumstances described by your correspondent.

At slack tide in Harwich harbor in Essex, and with a slack breeze, I leaned my belly against a stanchion on one ship, stretched with both hands across the narrow gap to a similar stanchion on the ship that was lying alongside, and pulled hard.

For perhaps half a minute there seemed to be no result,

but slowly the gap between them began to diminish until the two ships came quietly, and without fuss or noise, into contact. And, left alone, they remained in contact. Then, by reversing the process over a similar timescale, and substituting a push for a pull, the two ships returned to their starting positions. The process was remarkably simple.

The *QE2* is just a trifle larger than a Navy destroyer but I believe that the only difference would be in the timescale required to move the ship. Should your correspondent find an, admittedly unlikely, opportunity to try this experiment with such a large vessel I would advise that he takes care not to hold his breath while pulling.

KEN GREEN

There is no threshold force that needs to be overcome to move a ship in the absence of wind or current. In fact it is remarkably easy for an unassisted person to move a large ship. This can be explained in terms of kinetic energy (E) and momentum.

Consider a ship with a mass of 20,000 tons. If the ship is given a velocity of 0.5 inch per second then its energy, $E = \frac{1}{2} mv^2$ is about 1,000 joules. A thousand joules is a very modest amount of energy. It is the energy expended by a 120-pound man climbing up a 6-foot-high flight of stairs.

At 0.5 inch per second the ship's momentum (mass × velocity) = 2×10^5 newton-seconds. The 120-pound man can impart this to the ship by applying his full weight for 400 seconds. If he moves the ship by standing with full weight on one of the mooring lines, he will have descended by 6 feet by the time the ship is moving at 0.5 inches per second.

Actually when a ship is set in motion, a comparable mass of water is also set in motion at a comparable speed. Consequently the kinetic energy and momentum calculated above have been underestimated by a factor of two or so. However, the main conclusion stands: an unaided person can easily move a ship.

JOHN PONSONBY

The ship will move. Fluid forces don't have a limiting static friction. We can think of these frictional fluid forces as being directly proportional to the speed of the ship. They are close to zero when the speed is close to zero, and so on.

So push, and good luck!

MARCO VENTURINI AUTIERI

Pane Barrier

The two outer panes of a passenger aircraft cabin window have a tiny piece of cylindrical metal separating them. It is always near the base of the panes, not in the center, and is frequently surrounded by condensation. What purpose does it serve and what is it made of?

RITA BREITKOPF

Airline windows typically comprise three or more layers of glass (or acrylic) to provide insulation from the very cold atmosphere at altitude. The tiny silvery cylinder is really the edge of a small hole drilled in the middle layer to allow the pressure to equalize between the layers while minimizing convection.

The condensation around the hole is due to the inner air-space cooling. Ice often forms here. The position of the hole is chosen to maintain the best clear viewing area when condensation forms, to minimize the likelihood of a crack forming between the hole and the edge of the window, and to avoid excessive condensation pooling over the hole, which could freeze and block it.

FRED PARKINSON

Fasten Seatbelts

I recently flew back from vacation on a large airliner. During the flight we encountered severe turbulence. Food and drink went flying, overhead lockers opened, people were screaming and crying, and even the cabin crew were alarmed, crawling along the aisle to take refuge. The plane seemed to fall vertically for about 5 seconds. How much danger were we in? It felt as if the plane was going to fall out of the sky. Has that ever happened?

BRIAN JACKSON

The questioner has experienced the effects of clear air turbulence, or CAT. Pilots are unable to see CAT, and it can indeed cause an aircraft to crash, especially if encountered just after take-off or just before landing. Since 1981, there have been more than 350 reports of aircraft running into serious turbulence. It is also considered to be the leading cause of in-flight injuries: in the U.S., around 60 passengers are injured each year as a result of CAT, which is why passengers are often advised to keep their seatbelts fastened.

There are five major causes of CAT: jet streams; the wake of other aircraft; airflow over mountains; thermals and microbursts; and severe downdrafts associated with rain and thunder clouds. The encounter described here seems to have been caused by jet stream turbulence. On long flights, aircraft try to fly along the jet stream if they can, but since this is usually found at altitudes above 40,000 feet, planes often fly just below it, where there can be turbulent zones.

If an aircraft at cruising altitude encounters a downdraft, the wings lose lift and the aircraft drops suddenly. Anyone or anything not fastened down then hits the ceiling with varying degrees of force, and it is usually the flight attendants who suffer injuries as a result. When the aircraft flies out of the downdraft, the wings regain lift with a loud bang. Aircraft wings are designed to withstand 1.5 g of negative lift and 2.5 g positive—any more than this and damage will occur.

On March 5, 1966, CAT caused a major disaster. On a clear day the pilot of British Overseas Airways Corporation flight 911 decided to make a detour to give his passengers a sightseeing tour of Mount Fuji. The Boeing 707 broke apart as a result of the severe turbulence encountered as the aircraft flew too close to the mountain.

There are several other instances where aircraft have crashed on landing or take-off, often as a result of microbursts. Planes on approach have been flipped over and had insufficient altitude to take corrective action. For this reason, predictive wind-shear radar is now required by law on civil airliners in the U.S. Such radar can detect the water droplets usually associated with microbursts, and warn the pilot to turn away.

TERENCE HOLLINGWORTH

8 Best of the Rest

Family Line

My wife's parents have six children, the first three and the sixth of whom are girls. Four of these children have children of their own, seven in all, and all of them girls. I am aware that men in certain occupations are more likely to father girls, but in this family the seven grandchildren have four different fathers. Is this purely coincidence, or are there other factors at work?

MARK HIGGINS

Without knowing all the details, I will venture that coincidence seems likely. Any fetus has approximately a 1-in-2 chance of being female; the chance of seven children all being girls is therefore 1 in 2 to the seventh power, or 1 in 128. These are not particularly long odds, and if you consider that one would probably find several other combinations remarkable too (all boys, for example, or exactly alternating girls and boys), the odds that any given set of six grandchildren will exhibit a "remarkable" pattern is low enough that such an occurrence is, in fact, unremarkable.

This question stems from the human ability to find patterns in random data. It has been observed that people's expectations about what a random sequence looks like are actually quite different from a typical random sequence. With coin flips, for example, a person might cite HTHHTHTTHHHTTH as a typical sequence, but a truly typical random sequence might look more like HHHH-

HTHHTTHHHH: less alternation, longer runs of one value, and a ratio of heads to tails that is often far from 0.5 for short sequences. This means, conversely, that sequences that actually are random often appear not to be random to most people.

If the group eventually reached 15 grandchildren and all were girls, then I might raise an eyebrow and wonder whether there might be a nonrandom cause at work. But even here the probability is still low enough (1 in about 32,000) for it to have happened quite a few times by chance throughout history.

BEN HALLER

Cold Surface

I have heard that a common way to catch a cold is if somebody with the virus touches your hand before you touch your own nose or eyes. Apparently it can even be passed on via a third surface such as a door handle. How long can a cold virus or any other pathogen live on a surface? Does it depend on the surface and does moisture make a difference?

CORY CAULFELL

It depends on the surface. Cool, moist glass in the shade, for instance, might retain many kinds of rhinovirus or coronavirus for days.

Brass which is dry, sun-baked, and covered with verdigris and zinc compounds, on the other hand, might be germ-free within half an hour of being touched. Such compounds are bad for most microbes, so filthy lucre, especially coins made of copper alloys, is not nearly as horribly germy as one might expect.

By and large, rhinoviruses are the most common causes of colds. They are picornaviruses, which are generally only moderately stable. Desiccation and ultraviolet light in open sunshine should render most surfaces safe quite quickly. A cozy damp pocket handkerchief, though, might harbor the germs for days, unless it is infested with decay bacteria that digest viruses along with the nutritious secretions donated by the owner.

To avoid infection in a viral epidemic, it makes sense to avoid touching your face as much as possible and to wash your hands before doing so.

JON RICHFIELD

Killer Chemical

How does chlorine in swimming pools kill harmful organisms, and why is it the chemical of choice?

TOMMY KRONE

Chlorine is not the only member of the halogen group that can be used to disinfect water; iodine and bromine will also do the job, though not fluorine because it is too reactive. Chlorine is often chosen simply because it is cheap, readily available, and relatively easy to handle.

Disinfection relies on disrupting a harmful organism's metabolism or structure. That can be achieved by oxidation and nonoxidizing chemicals which have similar effects, as well as by nonchemical processes such as ultraviolet (including sunlight), X-rays, ultrasound, heat (as in pasteurization), variations in pH, and even storage to allow organisms to die naturally.

Chlorine gas consists of molecules of two chlorine atoms but no oxygen. When added to water, one of the atoms

forms a chloride ion. The other reacts with water to form hypochlorous acid, an oxidizing agent. Disinfection comes from the hypochlorous acid reacting with another molecule, most probably in the bacterial cell wall, in an oxidation-reduction reaction. If this happens enough times, the organism's repair mechanisms are overwhelmed and it dies. So concentration of disinfectant and the length of time pathogens are exposed to it are important factors.

Chlorine is available in many different chemical forms, such as chlorine gas, sodium hypochlorite powder (often used in home swimming pools), and chlorinated lime or bleaching powder. Some chemicals containing chlorine are not disinfectants because the chlorine in them, usually in the form of chloride, is completely reduced with no further oxidizing power. Sodium chloride is such a chemical, which is why water cannot be disinfected using a pinch of salt, and why pathogens can survive in seawater.

PHILIP JONES
Water Environment Consultants, Woking, Surrey, UK

Disinfection needs to be carried out under closely controlled pH conditions, ideally between 7 and 7.6. If the pH is too low—less than 6.8—there is a tendency for nitrogen compounds, especially urea (a common pool contaminant), to degrade via another route to chloramines. The worst of these is nitrogen trichloride, which irritates the eyes and creates the so-called chlorine smell associated with poorly run or overused swimming pools.

PHILIP STAINER
Lach Dennis Consultants, Haverhill, Suffolk, UK

Chlorination removes contamination immediately in the pool, whereas ultraviolet and ozone treatment work in the plant room. All these systems also use filters to remove organic matter. The less turbid the water, the lower the dose of chlorine needed to sterilize it. So a constant, low level of chlorine can be maintained in the water circulating between pool and plant room, and it can be altered as pool users and pollution levels vary.

Lois Vickers
Thanks also to Roger Cole of Proton Water Services

Pipe Dreams

During a conversation about playing the bagpipes at high altitude, I wondered what would happen to the sound of the bagpipes if they were played in the helium/oxygen mix used by deep sea divers, which distorts speech. Would the double reed chanter (the output part of the bagpipe that consists of a tube with holes and is played with the fingers) be affected in the same way as the single-reed drones (the output pipes confined to single notes)?

Roger Malton

The construction of a bagpipe allows a continuous supply of air to be maintained. A flexible bag is filled with air and acts as a reservoir. By squeezing the bag while a breath is taken, one can keep up the flow of air in both drone pipes and chanter.

The fundamental frequency of a resonating cavity, whether it is the voice or a resonating tube like a bagpipe chanter, is directly proportional to the speed of sound of the gas occupying the cavity. The speed of sound is proportional to the square root of T/M (where T is the absolute temperature of the gas and M is its molecular weight). Therefore the speed of sound is higher in gases with smaller molecular weights. For example, the speed of sound in air (where M = 28.964) at 32°F is 1,087 feet per second. And in helium (where M = 4.003) the speed is 2,924 feet per second. The resonance frequencies of the vocal tract are therefore almost

2.7 times higher for helium than for air and the pitch will be much higher than usual, rather like Donald Duck's.

The original question is, of course, the wrong one. It is difficult to imagine playing the Scottish bagpipe in the confines of a diving bell filled with the helium-oxygen mix. The question is more relevant to the Irish whistle which is easily portable and still satisfies a deep-seated human need for Celtic music.

I carried out an experiment by inhaling from a toy helium balloon with my brass Sindt D whistle 134 feet above sea-level, where the ambient temperature was 72°F. Once a stable note had been reached, the pitch jumped up almost exactly three semitones from D to F and remained in tune from then on. Although I had to blow harder to keep the notes constant I could play the first 12 bars of "Down by the Sally Gardens" without taking a breath, albeit slightly faster than usual. The air/helium mix I exhaled after taking the first breath of air returned the pitch to D sharp. However a pure D did not return for some time as residual helium was slowly cleared from my lungs. Residual lung volume accounts for about 25 percent of total lung volume, therefore the first breath of helium was probably about 75 percent mix, and the second approximately 18 percent, assuming that the gas inhaled from the balloon was pure.

TONY LAMONT

The pitch of both types of pipe in the bagpipes is determined by the effective length of the pipe (which is varied by opening holes in the chanter) and not by the reed. The reed adapts its frequency to the resonance set up in the pipe in which it sits. The frequencies of the modes of any pipe are proportional to the velocity of sound in the gas and, because this is much higher in helium than in air, the pitch of the bagpipes must rise.

I used to teach the physics of music to opera singers at a major music college, and they were always impressed when I took along a helium cylinder and had them fill their vocal cavities and lungs with it. When you do this you need to be careful to retain some carbon dioxide in your lungs because this stimulates the automatic breathing reflex. In the case of singers, the pitch does not in fact change, because it is determined by the vocal cords, not the pipe.

The resonances are not strong enough to dominate the heavy vocal cords and their di-muscular control. What does change is the frequency of every resonance of the vocal tract, and hence the tone color (actually, the formant) of the voice changes dramatically. The voice sounds higher because the color shifts to higher frequencies, not the actual pitch.

In practice, very few singers managed to hear much of their new voice, because they invariably laughed at the unfamiliar sound they produced and quickly expelled the helium.

John Elliot
UMIST, Manchester, UK

Yes, bagpipes do work with helium or helium mixtures—100 percent helium in the bag raises the pitch by about an octave and some retuning is required between chanter and drone.

Trials were done with bagpipes as a precursor to designing the heli racket, an instrument entered in the new musical instrument challenge run by BBC2 TV program *Local Heroes*. Helium and air were blown through a bagpipe chanter, and notes changed by varying the ratio of gases using a mixing valve (in this case, a bathroom tap), rather than finger holes. The instrument gave a passable televised rendition of "Twinkle Twinkle Little Star."

As an alternative to helium, mixtures of gases heavier than air, such as oxygen and neon, could be used to lower the pitch.

I can also report that changes of gas do nothing for tone quality.

Mark Williams

The resonant frequencies of all pipes and air chambers are directly proportional to the speed of sound. A helium/oxygen mixture will increase all the frequencies but carbon dioxide will have the reverse effect, so musicians beware.

Woodwind players know that they must avoid drinking fizzy drinks before performances. If you belch into the instrument as you play it, you fill it with carbon dioxide which has a lower sound velocity.

The instrument goes horribly flat and doesn't recover its pitch until all the carbon dioxide has been blown through. Changing from pure air to pure carbon dioxide would send the instrument about seven semitones flat.

LAURIE GRIFFITHS

In response to the question asking whether bagpipes sound better if played with a mixture of helium and oxygen instead of air: of course not, bagpipes already sound perfect.

JOE BOSWELL

Received Pronunciation

How do accents develop and change? More specifically, how do new accents form, such as those that arose in Australia and New Zealand? Presumably these are no more than 200 years old.

CIARON LINSTEAD

Accents and dialects develop and change for two distinct reasons, one phonetic, the other social. On the phonetic side, speech sounds change because of the way they are produced and perceived. Feel the position of your tongue against the roof of your mouth when you say the K sounds at the beginning of "key" and "car." The tongue makes contact farther forward in "key" than in "car," because it is anticipating its forward position for the vowel sound EE. This more forward position has led to changes in which K sounds became CH or SH or S sounds before EE or E vowels. The Latin word "centum" began with a K sound, but Italian "cento" begins with CH and French cent with S. These changes were among many that occurred as Latin evolved into modern Romance languages.

Phonetic changes don't happen continuously, though, because language is used to communicate. If your pronunciation is suddenly different from that of the people around you, you won't be understood. The communicative function of language provides a social brake on the phonetic causes of change. In any community, however, phonetic changes can take hold from one generation to the next. When communities are relatively isolated as Australia was from England during its development they may adopt different phonetic changes. This is how Australian and English pronunciations have diverged. Two hundred years is plenty of time for differences to develop.

This kind of divergence brings a more subtle social effect into play. The information you convey when you talk is not limited to the linguistic meaning of your words, but includes many things about yourself, such as regional origin or level of education. Speakers unconsciously (or consciously) tailor their speech to sound like the person they want to appear to be. This has an influence on the development of accents and their change: people adopt or reject specific sounds and sound changes to signal their identification with a particular community.

Bob Ladd
Professor of Linguistics
University of Edinburgh, UK

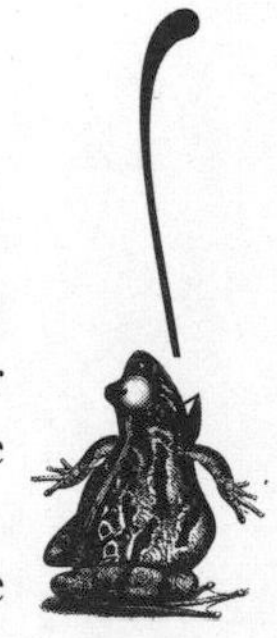

Starting with a relatively uniform speech community, minor variations in sounds may acquire greater or lesser prestige by association with individuals or groups who use them.

In Australia and New Zealand, the biggest divergence

from English "standard received pronunciation" is in the vowel system. In the early nineteenth century, there was a tendency in southern England, where many colonists came from, to pronounce the vowel in "bad" (known to phoneticians as RP Vowel No 4 or RP 4) in a more "closed" position (with the mouth less open) so it sounded more like "bed." Later, this trend was halted and partially reversed in England. Its southern base was relatively stagnant demographically compared with the booming North and Midlands, which kept the more open A in "bad." Today, the very closed version of vowel 4 is increasingly stigmatised as "hyper-posh" and causes surprise when heard in old 1940s newsreels.

By contrast, in Australia and New Zealand it flourished, perhaps cementing solidarity among the older settlers as against the later-arriving Poms, who had the more open vowel. The "closedness" was exaggerated further, causing potential confusion with RP 3, as in "bed." The latter vowel had to move over to make room for it, by becoming even more closed and sounding like "bid." The vowel in "bid" (RP 2) in turn had to become still more closed, to sound like RP 1 "bead," which in turn tended to become a diphthong, sounding something like "buyd." In New Zealand, the process was similar, except that RP 2 ("bid") was pushed into the center of the mouth, to sound like RP 10 as in "bud" or RP 12—the sound in the second syllable of "cupboard."

This phonetic musical chairs, caused by an initial point of imbalance in the system, is known to linguists as a "push-chain." RP 4 "pushed" the others to make room for itself.

There can also be a "pull-chain" in which a departing vowel leaves a slot into which a neighboring sound can expand. This also happened in Australia. Once RP 4 "bad" had moved over to "bed," the long back-vowel of "bard" RP 5 was free to drift to the front of the mouth, without fear of confusion.

Thus, RP 4 has "pulled" RP 5 after it. In support of this, in Australian soaps like *Neighbours*, older characters tend to have accents closer to received pronunciation, but younger ones have a pronunciation typical of the system outlined here.

STEVE TANNER

It is often assumed that accents in countries that see large-scale immigration will diverge from the accents of the settlers' original country. The reverse may be true, however. The original accent can remain in the country now occupied by immigrants, while the accent in the nation of origin develops along new lines. This has occurred in the development of American English.

The first permanent English immigrants to North America settled in Jamestown, Virginia, in 1607, while 13 years later the Pilgrim Fathers landed further north at what is now Plymouth, Massachusetts. The Cambridge Encyclopaedia of the English Language *by David Crystal tells us that these two settlements had different linguistic consequences for the development of American English. The Jamestown colonists came mainly from England's West Country and spoke with the characteristic burr of these counties. This pattern can still be heard in some of the communities of the Jamestown region, especially Tangier Island in Chesapeake Bay. Because of the relative isolation of this area, this "Tidewater" accent has changed only slightly in 400 years and is sometimes said to be the closest we will ever get to the sound of Shakespearean English.*

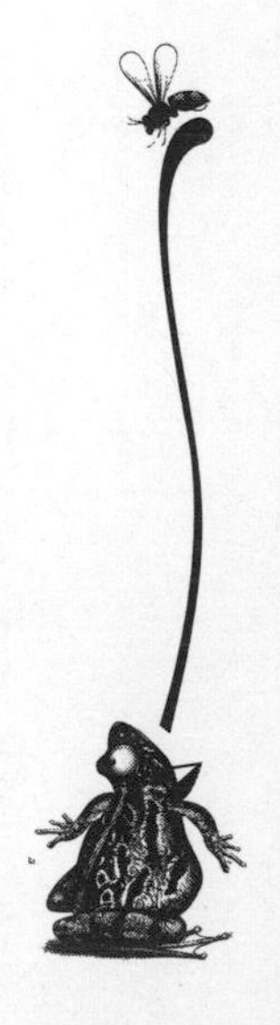

The Plymouth colonists, by contrast, came from eastern England. These accents dominated in what is now New England, and their speech patterns are still the main influence in this area. An outline of the development of English in all its forms can be found on the BBC website www.bbc.co.uk/routesof english.—Ed.

War Nuts

An entry in the local school logbook for the village of Nash in north Buckinghamshire, dated November 9, 1917, states "Letter of thanks received from the Director of Propellant Supplies for chestnuts gathered for the making of munitions."

JOHN HARRIS AND GREG DAVIES

We have had visions of chestnut shrapnel, but this can't be right, can it? So what were chestnuts used for and what is their link with propellants?—Ed.

This question was asked in the February 1987 issue of *Chemistry in Britain*, the Royal Society of Chemistry's monthly magazine. Subsequent correspondents gave the answer that the chestnuts were used in the First World War for the production of acetone which, in turn, was needed for the production of cordite, the smokeless powder used as propellant in small arms ammunition and artillery.

Smokeless powders such as cordite had changed the face of battle. They offered longer range than "black powder," better known as gunpowder. Producing only a faint blue-gray haze, they permitted machine guns to fire without obscuring the gunner's view and they permitted snipers to operate without revealing their position. Cordite is a mixture of the explosives guncotton (65 percent), nitroglycerine (30 percent), and petroleum jelly (5 percent), gelatinized with the aid of acetone before being worked into threads for use.

Prewar techniques for large-scale acetone manufacture were inadequate to meet the demand during the First World War. As minister of munitions, David Lloyd George appointed Chaim Weizmann, a chemist who had emigrated from mainland Europe in 1904, to increase acetone production using a process of his own invention involving the bacterial fermentation of maize starch. Factories at Poole in

Dorset and King's Lynn in Norfolk produced up to 90,000 gallons of acetone a year. When supplies of maize ran short, it was supplemented as a source of starch with horse chestnuts, collected by schoolchildren. Since the factory locations were withheld for the sake of security, the schools addressed their parcels to government offices in London, but workers at the General Post Office apparently knew to send them directly to the factories.

This is how Lloyd George's acetone problem left traces in school record books. It also left, in his own words, "a permanent mark on the map of the world." Lloyd George was so grateful to Weizmann, an ardent Zionist, that on becoming prime minister he gave Weizmann direct access to the foreign secretary, A. J. Balfour. The result was the famous and controversial "Balfour Declaration'"of November 2, 1917, stating that the British government viewed with qualified favor "the establishment in Palestine of a national home for the Jewish people." When the state of Israel came into existence, Weizmann was elected as its first president in 1948 and held that position until his death in 1952.

MICHAEL GOODE

Tony Cross, who is the curator of the Curtis Museum in Alton, Hampshire, drew our attention to similar school records from his own area and to an explanation he received for them from the Imperial War Museum. The answer below summarizes that explanation.—Ed.

During the First World War some 258 million shells were used by the British Army and the Royal Navy. The basic propellant used to fire these shells, and for a whole host of other military purposes, was cordite. The solvents used in manufacturing cordite were acetone and ether-alcohol.

Acetone was produced almost entirely by the destructive distillation of wood, for which the world market was dominated by the great timber-growing countries. Before the war acetone was mainly imported from the United States. In 1913 a modern factory was established in the Forest of Dean but, by the outbreak of war in August 1914, the stocks of acetone for military use stood at only 3,200 tons. It was soon apparent that production would not meet the rapidly growing demand. When it was discovered that acetone could be produced from potatoes and maize, new factories were erected to undertake this work.

By 1917, however, the German submarine offensive in the Atlantic had caused a shortage of freight which threatened to cut off supplies of North American maize. With the possibility of a serious maize shortage, experiments began to find a substitute and it was discovered that the horse chestnut could be used as an alternative in acetone production. Vast quantities of horse chestnuts were collected, but only 3,000 tons reached the King's Lynn plant. Collection was restricted by transport difficulties, and letters in *The Times* tell of piles of rotting horse chestnuts at railway stations.

After initial production difficulties the King's Lynn factory began production of acetone from horse chestnuts in April 1918. Work was hampered by the fact that the horse chestnut was a poor-quality material from which to produce acetone. The plant was eventually closed in July 1918.

Index